Hands-On Mobile and Embedded Development with Qt 5

Build apps for Android, iOS, and Raspberry Pi with C++ and Qt

Lorn Potter

BIRMINGHAM - MUMBAI

Hands-On Mobile and Embedded Development with Qt 5

Commissioning Editor: Kunal Chaudhari
Acquisition Editor: Devanshi Doshi
Content Development Editor: Pranay Fereira
Technical Editor: Jane D'souza
Copy Editor: Safis Editing
Project Coordinator: Kinjal Bari
Proofreader: Safis Editing
Indexer: Manju Arasan
Graphics: Alishon Mendonsa
Production Coordinator: Jyoti Chauhan

First published: April 2019

Production reference: 1300419

Published by Packt Publishing Ltd.
Livery Place
35 Livery Street
Birmingham
B3 2PB, UK.

ISBN 978-1-78961-481-7

www.packtpub.com

To Qt, its founders, the developers who have worked on it, and its community of users.

`mapt.io`

Mapt is an online digital library that gives you full access to over 5,000 books and videos, as well as industry leading tools to help you plan your personal development and advance your career. For more information, please visit our website.

Why subscribe?

- Spend less time learning and more time coding with practical eBooks and Videos from over 4,000 industry professionals

- Improve your learning with Skill Plans built especially for you

- Get a free eBook or video every month

- Mapt is fully searchable

- Copy and paste, print, and bookmark content

Packt.com

Did you know that Packt offers eBook versions of every book published, with PDF and ePub files available? You can upgrade to the eBook version at `www.packt.com` and as a print book customer, you are entitled to a discount on the eBook copy. Get in touch with us at `customercare@packtpub.com` for more details.

At `www.packt.com`, you can also read a collection of free technical articles, sign up for a range of free newsletters, and receive exclusive discounts and offers on Packt books and eBooks.

Foreword

During Easter 2019, I got a message from Lorn Potter, asking if I could write a foreword for his book about mobile and embedded development with Qt 5. I said, "sure, I'd be happy to," knowing what a skilled guy he is. Then he told me that the deadline was in a few hours. Challenge accepted!

I think he probably chose me because I've known him for a very long time, and maybe also because I am one of the two founders of Trolltech and also one of the first two Qt designers and developers.

Back in the stone ages, when Trolltech re-licensed Qt for Windows, to make it available for the development of open source software, we got an email from a guy called Lorn Potter. He wrote to us: "Trolltech, if you were a woman, I'd... I'd... Well, I'd take you out for dinner and a movie!". A bit later we worked on putting together some marketing material and someone thought it would be cool to use that quote. Marketing asked for permission to use the quote and contact with Lorn was established. Another story is that that quote created quite a controversy internally and was later removed.

Lorn was the creator of a desktop app called Gutenbrowser and quite a rock star in the open source community. When we needed a community liaison for our Qtopia product in mid-2003, Lorn was offered the job and took it. He then moved from the US to Brisbane, Australia, to work at our office there. Qtopia was basically a complete platform and application suite for mobile devices and was built with Qt.

Lorn was heavily involved with the open source developed version of Qtopia. Other duties he had at Trolltech included creating the Qtopia Greenphone SDK for our only hardware product, the Trolltech Greenphone:

He also ported Qtopia to run on other devices, such as Nokia's N770, N800, N900, Siemens SIMpad SL4, and Openmoko's Neo 1973 phone.

Lorn has decades of experience in using Qt for mobile and embedded development. Here is a picture of his collection of devices running Qt, just to give you an idea:

I haven't read this book, so I cannot guarantee its quality, but I can guarantee the quality of Lorn as a communicator and a very talented developer. You are in the best of hands.

Now read on and use Qt, my baby who has grown up to be a very powerful adult.

Eirik Chambe-Eng, Co-founder, Trolltech

Contributors

About the author

Lorn Potter is a software developer, specializing in Qt and QML on mobile and embedded devices with his company, llornkcor technologies. He has worked for Trolltech, Nokia, Canonical, and was a freelance contractor for Jolla, Intopalo, and the Qt Company. He is the official maintainer of Qt Sensors, for which he developed the QSensorGestures API. He maintains the unsupported QtSystemInfo for the open source Qt Project and also works on Qt Bearer Management and Qt for WebAssembly. He has written blogs and articles for the Linux Journal. He started his career in tech as Trolltech's Qtopia Community Liaison. He currently resides in Australia and spends his spare time recording electronic psybient music for the project Brog on his website.

About the reviewer

Mickael Minarie is a software developer who graduated from University of Clermont Ferrand with a BSc in embedded systems, and from Robert Gordon University with a BSc in computer science. He has worked freelance, developing some programs in C++/Qt for embedded systems and other programs related to photos and videos.

He lives now in France, but stayed for some years in the UK and Canada. He is enthusiastic about the techniques and history of photography and audio recording, and he now runs a YouTube channel discussing these subjects (in French only, for now). Mickael has also created some small projects for companies using Raspberry Pi and Qt in connection with photography.

Packt is searching for authors like you

If you're interested in becoming an author for Packt, please visit `authors.packtpub.com` and apply today. We have worked with thousands of developers and tech professionals, just like you, to help them share their insight with the global tech community. You can make a general application, apply for a specific hot topic that we are recruiting an author for, or submit your own idea.

Table of Contents

Section 2: Networking, Connectivity, Sensors, and Automation

Preface

Qt is everywhere these days. From your typical home computer to cloud servers, mobile phones, machine automation, coffee machines, medical devices, and embedded devices of all kinds, even in some of the classiest automobiles. It might even be somewhere in space, too! Even watches run Qt these days.

The **Internet of Things (IoT)** and home automation are big buzzwords that Qt is also a part of. As I like to say, there's no IoT without sensors! Would I write a book on Qt development without mentioning sensors? No. So, we will also dive into Qt Sensors as well.

There's no shortage of target devices these days, that's for sure. But in this book, we will only specifically target mobile phone platforms and Raspberry Pi to demonstrate some of the embedded features of Qt. Because Qt is cross-platform, you should still come away with knowledge on how to target coffee makers, too!

One consideration to make is which UI framework is going to be used. Using OpenGL is a viable option for mobile devices, especially if there is hardware support. I deliberately skip discussing OpenGL, as that is quite complicated and would be a complete book in itself. It is an option, however, and is available through the use of Qt frameworks.

Who this book is for

This book is aimed at developers who are interested in developing cross-platform applications with Qt for use on mobile and embedded devices. Readers should have prior knowledge of C++ and be familiar, running commands from a command-line interface.

What this book covers

Chapter 1, *Standard Qt Widgets*, covers standard UI elements and dynamic layouts to teach the reader how to handle orientation changes.

Chapter 2, *Fluid UI with Qt Quick*, outlines standard QML elements, charts, data visualization, and animation.

Chapter 3, *Graphical and Special Effects*, looks at QML particles and graphical effects.

Chapter 4, *Input and Touch,* teaches the readers how to create and use virtual keyboards, and touch and voice input.

Chapter 5, *Qt Network for Communication,* talks the reader through network operations, sockets, and sessions.

Chapter 6, *Connectivity with Qt Bluetooth LE,* goes over Bluetooth LE device discovery, setting up a service, and manipulating characteristics.

Chapter 7, *Machines Talking,* discusses sensors, WebSockets, and MQTT automation.

Chapter 8, *Where Am I? Location and Positioning,* looks at GPS location, positioning, and maps.

Chapter 9, *Sounds and Visions – Qt Multimedia,* covers 3D audio, FM radio, and recording and playing video.

Chapter 10, *Remote Databases with Qt SQL,* outlines, remote use of SQLite and MySQL databases.

Chapter 11, *Enabling In-App Purchases with Qt Purchasing,* discusses adding in-app purchases to your apps.

Chapter 12, *Cross Compiling and Remote Debugging,* looks at cross-compiling, and connecting to and debugging on a remote device.

Chapter 13, *Deploying to Mobile and Embedded,* examines the setting up of an application and completion of app store procedures.

Chapter 14, *Universal Platform for Mobiles and Embedded Devices,* looks at running Qt apps in a web browser.

Chapter 15, *Building a Linux System,* covers setting up and building a complete Linux embedded system.

To get the most out of this book

This book assumes you have used C++, are familiar with QML, can use Git to download source code, and can type commands into a command-line interface. You should also be accustomed to using GDB debugger.

It also assumes you have a mobile or embedded device, such as a phone or Raspberry Pi.

Download the example code files

You can download the example code files for this book from your account at `www.packt.com`. If you purchased this book elsewhere, you can visit `www.packt.com/support` and register to have the files emailed directly to you.

You can download the code files by following these steps:

1. Log in or register at `www.packt.com`.
2. Select the **SUPPORT** tab.
3. Click on **Code Downloads & Errata**.
4. Enter the name of the book in the **Search** box and follow the onscreen instructions.

Once the file is downloaded, please make sure that you unzip or extract the folder using the latest version of:

- WinRAR/7-Zip for Windows
- Zipeg/iZip/UnRarX for Mac
- 7-Zip/PeaZip for Linux

The code bundle for the book is also hosted on GitHub at `https://github.com/PacktPublishing/Hands-On-Mobile-and-Embedded-Development-with-Qt-5/tree/master`. In case there's an update to the code, it will be updated on the existing GitHub repository.

We also have other code bundles from our rich catalog of books and videos available at `https://github.com/PacktPublishing/`. Check them out!

Download the color images

We also provide a PDF file that has color images of the screenshots/diagrams used in this book. You can download it here: `https://www.packtpub.com/sites/default/files/downloads/9781789614817_ColorImages.pdf`.

Conventions used

There are a number of text conventions used throughout this book.

`CodeInText`: Indicates code words in text, database table names, folder names, filenames, file extensions, pathnames, dummy URLs, user input, and Twitter handles. Here is an example: "On *iOS*, you need to edit the `plist.info` file."

A block of code is set as follows:

```
if (!QTouchScreen::devices().isEmpty()) {
    qApp->setStyleSheet("QButton {padding: 10px;}");
}
```

Bold: Indicates a new term, an important word, or words that you see onscreen. For example, words in menus or dialog boxes appear in the text like this. Here is an example: "Select the **Projects** icon on the left side of **Qt Creator**, then pick which target platform you want like **Qt 5.12.0 for iOS**"

Warnings or important notes appear like this.

Tips and tricks appear like this.

Get in touch

Feedback from our readers is always welcome.

General feedback: If you have questions about any aspect of this book, mention the book title in the subject of your message and email us at `customercare@packtpub.com`.

Errata: Although we have taken every care to ensure the accuracy of our content, mistakes do happen. If you have found a mistake in this book, we would be grateful if you would report this to us. Please visit `www.packt.com/submit-errata`, selecting your book, clicking on the Errata Submission Form link, and entering the details.

Piracy: If you come across any illegal copies of our works in any form on the Internet, we would be grateful if you would provide us with the location address or website name. Please contact us at `copyright@packt.com` with a link to the material.

If you are interested in becoming an author: If there is a topic that you have expertise in and you are interested in either writing or contributing to a book, please visit `authors.packtpub.com`.

Reviews

Please leave a review. Once you have read and used this book, why not leave a review on the site that you purchased it from? Potential readers can then see and use your unbiased opinion to make purchase decisions, we at Packt can understand what you think about our products, and our authors can see your feedback on their book. Thank you!

For more information about Packt, please visit `packt.com`.

Section 1: Making Great UIs

Qt provides several ways for a developer to create great looking applications, from the more standard looking Qt Widgets to Qt Quick, with its endlessly creative fluid animation and sparkling effects, to Qt OpenGL, which brings immersive gaming into the Qt mobile world. In this section readers will learn how to use Qt's various UI frameworks to create dynamic and flexible user interfaces for mobile and embedded devices, targeting platforms, such as iOS, Android for mobile, and the Raspberry Pi.

This section comprises of the following chapters:

- Chapter 1, *Standard Qt Widgets*
- Chapter 2, *Fluid UI with Qt Quick*
- Chapter 3, *Graphical and Special Effects*
- Chapter 4, *Input and Touch*

Standard Qt Widgets 1

Qt Widgets are not the new kid on the block, but they still do have their place in applications that target mobile and embedded devices. They are well formed, predictable and have standard UI elements.

Recognizable UI elements are found in Qt Widgets and work great on laptops, which are simply mobile desktops. In this chapter, you will learn to design standard looking applications. Basic widgets such as menus, icons, and lists will be discussed with an emphasis on how to constrain the user interface to medium and small-sized displays. Topics we will discuss include how to use Qt's dynamic layouts to handle orientation changes. Classes such as QGraphicsScene, QGraphicsView, and QGraphicsItem will be used. Layout API such as QVBoxLayout, QGridLayout, and QStackedLayout will be discussed.

In this chapter we will cover:

- Using Qt Creator and Qt Widgets to create a mobile app and run on the device
- Differences between desktop and mobile apps including screen size, memory, gestures
- Using Qt Widgets in dynamic layouts for easy screen size and orientation changes
- Using QGraphicsView for graphical apps

Hello mobile!

So you want to develop apps for mobile and embedded devices using Qt. Excellent choice, as Qt was made for cross-platform development. To get you started, we will run through the basic procedure of using Qt Creator to create, build and run an application. We will briefly examine different aspects to consider when creating mobile and embedded apps, such as how to use Qt Creator to add a menu. Adding a `QWidget` in the designer is not that difficult, and I will show you how.

Qt has a long history of running on mobile devices, starting with Qt Embedded, which was initially released in 2000. Qt Embedded was the base framework for the UI Qtopia, which was initially released on the Sharp Zaurus on the SL-5000D developer edition.

These days, you can develop an application using Qt and sell it in the iOS App Store, Android Google Play store, or other Linux mobile phones. Qt apps run on TVs and you can even see them running on entertainment systems in cars and planes. It runs on medical devices as well as industrial automation machines on factory floors.

There are considerations for using Qt on mobile and embedded devices such as memory constraints and display size constraints. Mobiles have touchscreens, and embedded devices might not have screens at all.

When you install Qt, you can use the Qt Creator IDE to edit, build and run your code. It's free and open source, so you can even customize it. I once had a patch that customized Qt Creator in a way that would allow me to print out all the keyboard commands that is was using, so I could have a quick reference sheet. Let's take a quick look at Qt Creator, that was once known as Workbench.

Qt Creator

We are not going to go into any great detail about Qt Creator, but I thought I should mention it to demonstrate how we could go about using it to develop a cross-platform `QWidget` based application that runs on a desktop and mobile platform. Some differences between the two will be discussed. We'll then demonstrate how using dynamic layouts can help you target many different screen sizes and handle device orientation changes. You might already be familiar with Qt Creator, so we will refresh your memory.

Basic Qt Creator procedure

The basic procedure for cross-compiling and building apps that run on a mobile device are straight forward after you get set up. The procedure that we would hypothetically follow is:

1. **File | New File or Project... | Qt Widgets Application**, click the **Choose...** button
2. Write some amazing code
3. Select the **Projects** icon on the left side of **Qt Creator**, then pick which target platform you want like **Qt 5.12.0 for iOS**
4. Hit *Ctrl + B*, or *Command + B* to build
5. Hit *Ctrl + R*, or *Command + R* to run
6. Hit *F5*, or *Command + Y* to debug

For this first chapter, we will use Qt Widgets, which are UI elements that are more closely aligned to traditional desktop computer applications. They are still useful for mobile and embedded devices.

Qt Designer

Qt Creator comes with a design tool named Qt Designer. When you create a new template application, you will see a list of files on the left. It will open your application form in Qt Designer when you click on any `.ui` file.

The source code can be found on the Git repository under the `Chapter01-a` directory, in the `cp1` branch.

Navigate to **Forms | mainwindow.ui** and double click on that. This will open the UI file in Qt Creators Designer. A UI file is just a text file in the form of XML, and you can edit that file directly if you choose. The following image shows how it looks when opened in Qt Designer:

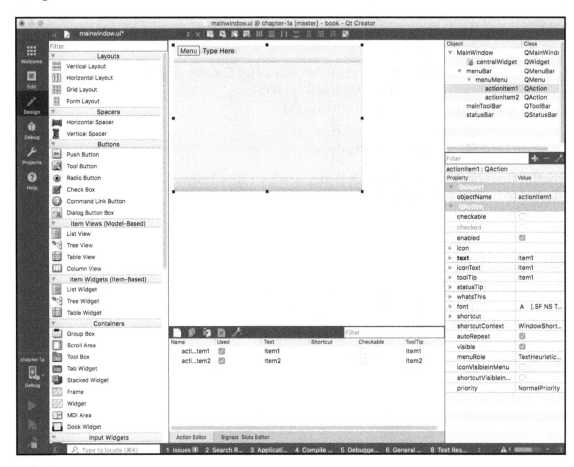

Let's start with something just about every desktop application has—a **Menu**. Your mobile or embedded application might even need a **Menu**. As you can see, there is a template **Menu** that the Qt app wizard has produced for us. We need to customize this to make it usable. We can add some sub-menu items to demonstrate basic Qt Creator functionality.

Add a QMenu

Click on the application form where it says **Menu** to add menu items. Type in something like Item1, hit *Enter*. Add another menu item, as demonstrated in the following image:

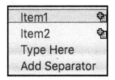

If you were to build this now, you would have an empty application with a **Menu**, so let's add more to demonstrate how to add widgets from the list of widgets that is on the left side of Qt Creator.

Add QListView

Our UI form needs some content. We will build and run it for the desktop, then build and run it for the mobile simulator to compare the two. The procedure here is easy as drag and drop.

On the left side of Qt Creator is a list of **Widgets**, **Layouts** and **Spacers** that you can simply drag and drop to place onto the template form and create your masterpiece Qt application. Let's get started:

1. Drag a **ListView** and drop it on the form.

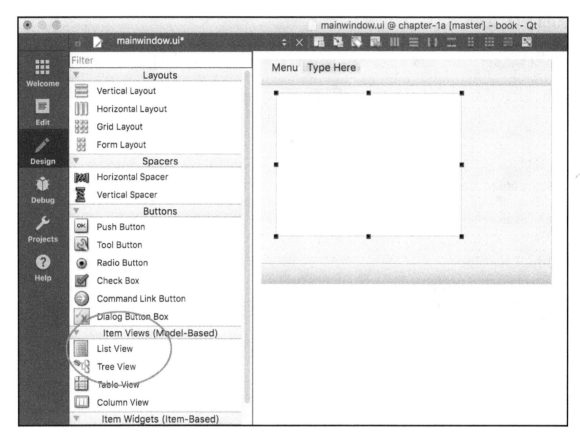

2. Select **Desktop** kit and build and run it by hitting the Run button. Qt Creator can build and run your application in the same step if you have made any changes to the form or source code. When you run it, the application should look similar to this image:

That's all fine and dandy, but it is not running on anything small like a phone.

Qt Creator comes with iOS and Android simulators, which you can use to see how your application will run on a small screened device. It is not an emulator, which is to say it does not try to emulate the device hardware, but simply simulates the machine. In effect, Qt Creator is building and running the target architectures.

3. Now select **iOS Simulator** kit, or `Android` from the **Projects** tool in green, as seen in the following image:

4. Build and run it, which will start it in the simulator.

Here is this app running on the iOS simulator:

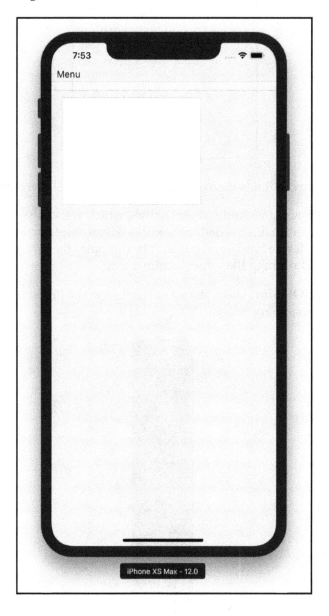

There you go! You made a mobile app! Feels good, doesn't it? As you see, it looks slightly different in the simulator.

Going smaller, handling screen sizes

Porting applications which were developed for the desktop to run on smaller mobile devices can be a daunting task, depending on the application. Even creating new apps for mobiles, a few considerations need to be made, such as differences in screen resolution, memory constraints, and handling orientation changes. Touch screens add another fantastic way to offer touch gestures and can be challenging due to the differences in the size of a finger as point compared to a mouse pointer. Then there are sensors, GPS and networking to contemplate!

Screen resolution

As you can see in the previous images in the *Add QListView* section, the application paradigms are fairly different between desktop and mobile phones. When you move to an even smaller display, things start to get tricky in regards to fitting everything on the screen.

Luckily, there are Qt Widgets that can help. The C++ classes `QScrollArea`, `QStackedWidget` and `QTabbedWidget` can show content more appropriately. Delegating your on-screen widgets to different pages will allow your users the same ease-of-navigation which a desktop application allows.

There might also be an issue on mobile devices while using `QMenu`. They can be long, unruly and have a menu tree that drills down too deeply for a small screen. Here's a menu which works well on a desktop:

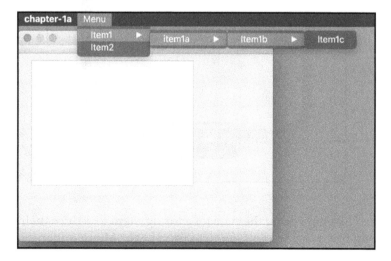

On a mobile device, the last items on this menu become unreachable and unusable. We need to redesign this!

Menus can be fixed by eliminating them or refactoring them to reduce their depth, or by using something like a `QStackedWidget` to present the **Menu** options.

Qt has support for high (**Dots Per Inch**) **DPI** displays. Newer versions of Qt automatically compensate for differences between high DPI and low DPI displays for iOS and the Wayland display server protocol. For Android, the environmental variable QT_AUTO_SCALE_FACTOR needs to be set to true. To test different scale factors, set QT_SCALE_FACTOR, which works best with an integer, typically 1 or 2.

Let's run through a few examples of widgets and how they can be better used on differing screens:

- Widgets like QScrollBar can be increased in size to better accommodate a finger as pointer, or better yet be hidden and use the widget itself to scroll. The UI usually needs to be simplified.
- Long QListViews can present some challenges. You can try to filter or add a search feature for such long lists to make the data more accessible and eye pleasing on smaller displays.
- Even QStackedWidget or QTabbedWidget can be too big. Don't make the user flick left or right more than a few pages. Anything more can be cumbersome and annoying for the user to be flicking endlessly to get at content.
- QStyleSheets are a good way to scale for smaller display's, allowing the developer to specify customizations to any widget. You can increase the padding and margins to make it easier for finger touch input. You can either set a style on a specific widget or apply it to the entire QApplication for a certain class of widget.

```
qApp->setStyleSheet("QButton {padding: 10px;}");
```

or for one particular widget it would be:

```
myButton->setStyleSheet("padding: 10px;");
```

Let's apply this only when there is a touch screen available on the device. It will make the button slightly bigger and easier to hit with a finger:

```
if (!QTouchScreen::devices().isEmpty()) {
    qApp->setStyleSheet("QButton {padding: 10px;}");
}
```

 If you set one style with a style sheet, you will most likely need to customize the other properties and sub-controls as well. Applying one style sheet removes the default style.

Of course, it is also easy to set a style sheet in Qt Designer, just right click on the target widget and select, **Change styleSheet** from the context menu. As seen here on the Apple Mac:

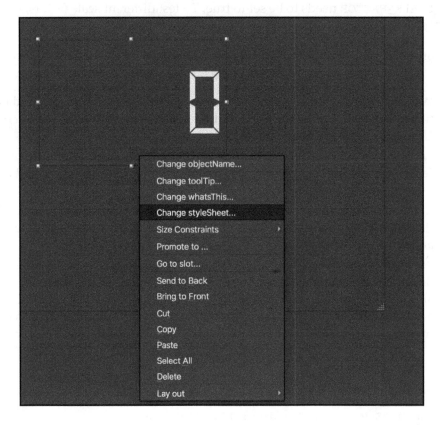

Mobile phones and embedded devices have smaller displays, and they also have less RAM and storage.

Memory and storage

Mobile phones and embedded devices usually have less memory than desktop machines. Especially for embedded devices both RAM and storage are limited.

The amount of storage space used can be lowered by optimizing images, compressing if needed. If different screen sizes are not used, the images can by manually resized instead of scaling at runtime.

There are also heap vs stack considerations which generally always pass arguments into functions by reference by using the & (ampersand) operator. You will notice this in the majority of Qt code.

Compiler optimizations can greatly effect both performance and the size of executables. In general, Qt's qmake mkspec build files are fairly good at using the correct optimizations.

If storage space is a critical consideration, then building Qt yourself is a good idea. Configuring Qt using the −no−feature−* to configure out any Qt features you might not need is a good way to reduce it's footprint. For example, if a device has one static Ethernet cable connection and does not need network bearer management, simply configure Qt using −no−feature−bearermanagement. If you know you are not using SQL why ship those storage using libraries? Running configure with −−list−features argument will list all the features available.

Orientation

Mobile devices move around (whodathunkit?) and sometimes it is better to view a particular app in landscape mode instead of portrait. On Android and iOS, responding to orientation changes are built in and occurs by default according to the users configuration. One thing you might need to do, is actually disable the orientation change.

On **iOS**, you need to edit the plist.info file. For the key UISupportedInterfaceOrientations, you need to add the following:

```
<array><string>UIInterfaceOrientationLandscapeLeft</string></array>
```

On **Android**, edit the AndroidManifest.xml file android:screenOrientation="landscape"

If a picture frame device has a custom-built operating system, it might need it's photo viewing app to respond when the user switches orientations. That's where Qt Sensors can help out. More on that later in the first section of Chapter 7, *Machines Talking*.

Gestures

Touchscreen gestures are another way mobiles are different to desktops. Multi-touch screens have revolutionized the device world. QPanGesture, QPinchGesture and QSwipeGesture can be used to great effect on these devices, and Qt Quick has components build for this type of thing—Flickable, SwipeView, PinchArea and others. More on Qt Quick later.

To use QGestures, first create a QList of containing the gestures you want to handle, and call the grabGesture function for the target widget.

```
QList<Qt::GestureType> gestures;
gestures << Qt::PanGesture;
gestures << Qt::PinchGesture;
gestures << Qt::SwipeGesture;
for (Qt::GestureType gesture : gestures)
    someWidget->grabGesture(gesture);
```

You will need to derive from and then override the widgets event loop to handle when the event happens.

```
bool SomeWidget::event(QEvent *event)
{
    if (event->type() == QEvent::Gesture)
        return handleGesture(static_cast<QGestureEvent *>(event));
    return QWidget::event(event);
}
```

To do something useful with the gesture, we could handle it like this:

```
if (QGesture *swipe = event->gesture(Qt::SwipeGesture)) {
    if (swipe->state() == Qt::GestureFinished) {
        switch (gesture->horizontalDirection()) {
            case QSwipeGesture::Left:
            break;
            case QSwipeGesture::Right:
            break;
            case QSwipeGesture::Up:
            break;
            case QSwipeGesture::Down:
            break;
        }
    }
}
```

Devices with sensors also have access to `QSensorGesture`, which enable motion gestures such as shake. More on that later, in `Chapter 7`, *Machines Talking*.

Dynamic layouts

Considering that mobile phones come in all shapes and sizes, it would be ridiculous to need to provide a different package for every different screen resolution. Hence we will use dynamic layouts.

The source code can be found on the Git repository under the `Chapter01-layouts` directory, in the `cp1` branch.

Qt Widgets have support for this using classes such as `QVBoxLayout` and `QGridLayout`.

Qt Creator's designer is the easiest way to develop dynamic layouts. Let's go through how we can do that.

To set up a layout, we position a widget on the application form, and press *Command* or *Control* on the keyboard while selecting the widgets that we want to put in a layout. Here are two `QPushButtons` selected for use:

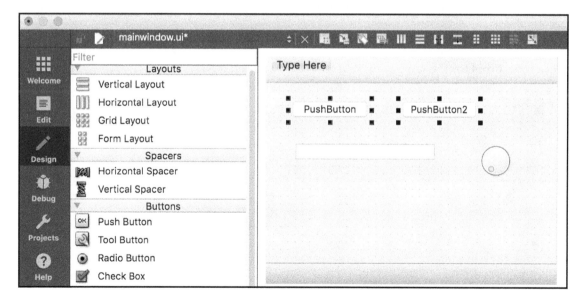

Next, click on the **Horizontal Layout** icon highlighted here:

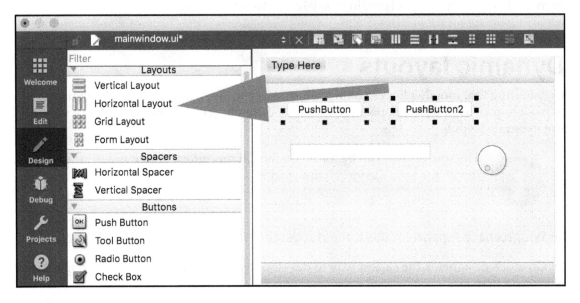

You will then see the two widgets enclosed by a thin red box as in the following screenshot:

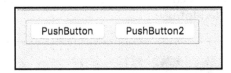

Now repeat this for the remaining widgets.

To make the widgets expand and resize with the form, click on the background and select **Grid Layout:**

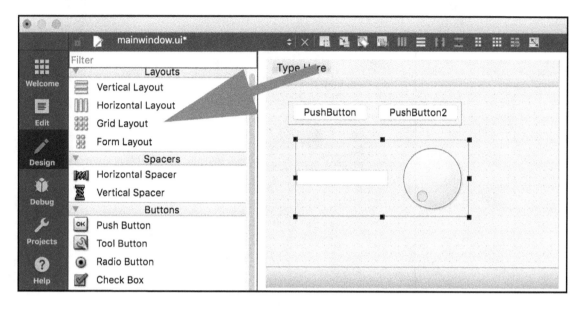

Save and build this, and this app will now be able to resize for orientation changes as well as being able to work on different sized screens. Notice how this looks like in portrait (vertical) orientation:

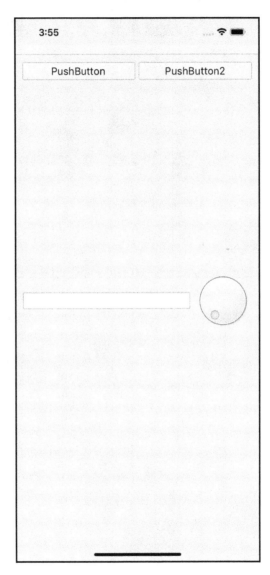

Also note how this same application looks in landscape (horizontal) orientation:

As you can see, this application can change with orientation changes, but all the widgets are visible and usable. Using QSpacer will help guide the widgets and layouts positioning. They can push the widgets together, apart, or hold some to one side or another.

Of course, layouts can be used without touching Qt Designer. For example the following code:

```
QPushButton *button = new QPushButton(this);
QPushButton *button2 = new QPushButton(this);
QBoxLayout *boxLayout = new QVBoxLayout;
boxLayout->addWidget(button);
boxLayout->addWidget(button2);
QHBoxLayout *horizontalLayout = new QHBoxLayout;
horizontalLayout->setLayout(boxLayout);
```

QLayout and friends are the key to writing a cross-platform application that can accommodate the myriad screen resolutions and dynamically changing orientations of the target devices.

Graphics view

QGraphicsView, QGraphicScene and QGraphicsItem provide a way for applications based on Qt Widgets to show 2D graphics.

> The source code can be found on the Git repository under the `Chapter01-graphicsview` directory, in the `cp1` branch.

Every `QGraphicsView` needs a `QGraphicsScene`. Every `QGraphicsScene` needs one or more `QGraphicsItem`.

`QGraphicsItem` can be any of the following:

- `QGraphicsEllipseItem`
- `QGraphicsLineItem`
- `QGraphicsLineItem`
- `QGraphicsPathItem`
- `QGraphicsPixmapItem`
- `QGraphicsPolygonItem`
- `QGraphicsRectItem`
- `QGraphicsSimpleTextItem`
- `QGraphicsTextItem`

Qt Designer has support for adding `QGraphicsView`. You can follow these steps to do so:

1. Drag the `QGraphicsView` to a new application form and fill the form with a `QGridLayout` like we did before.

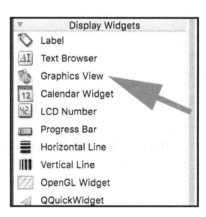

2. Implement a `QGraphicsScene` in the source code and add it to the `QGraphicsView`

```
QGraphicsScene *gScene = new QGraphicsScene(this);
ui->graphicsView->setScene(gScene);
```

3. Define a rectangle which will be the extent of the `Scene`. Here it is smaller than the size of the graphics view so we can go on and define some collision detection.

```
gScene->setSceneRect(-50, -50, 120, 120);
```

4. Create a red rectangle to show the bounding rectangle. To make it a red color, create a `QPen` which will be used to paint the rectangle and then add the rectangle to the `Scene`.

```
QPen pen = QPen(Qt::red);
gScene->addRect(gScene->sceneRect(), pen);
```

5. Build and run the application. You will notice an app with a red bordered square on it.

As mentioned before, `QGraphicsView` shows `QGraphicsItems`. If we want to add some collision detection we need to subclass `QGraphicsSimpleTextItem`.

The header file for this is as follows:

```
#include <QGraphicsScene>
#include <QGraphicsSimpleTextItem>
#include <QGraphicsItem>
#include <QPainter>
class TextGraphic :public QGraphicsSimpleTextItem
{
public:
    TextGraphic(const QString &text);
    void paint(QPainter *painter, const QStyleOptionGraphicsItem *option,
QWidget *widget);
    QString simpleText;
};
```

This custom class derived from `QGraphicsSimpleTextItem` will reimplement the `paint(..)` function, and use the `collidingItems(...)` function of `scene` to detect when something collides with our text object. Normally, `collidingItems` will return a `QList` of `QGraphicsItems`, but here it is just used to detect if any items are colliding.

Since this class holds only one item, it is known which item it is. If a collision is detected, the text changes. We don't need to check if the item's text is different before we change it, as the parent class's `setText(...)` does that for us.

```
TextGraphic::TextGraphic(const QString &text)
  : QGraphicsSimpleTextItem(text),
      simpleText(text)
{
}

void TextGraphic::paint(QPainter *painter, const QStyleOptionGraphicsItem
*option, QWidget *widget)
{
    if (scene()->collidingItems(this).isEmpty())
        QGraphicsSimpleTextItem::setText("BOOM!");
    else
        QGraphicsSimpleTextItem::setText(simpleText);

    QGraphicsSimpleTextItem::paint(painter, option, widget);
}
```

Now create our `TextGraphic` object and add it to the `Scene` to use.

```
TextGraphic *text = new TextGraphic(QStringLiteral("Qt Mobile!"));
gScene->addItem(text);
```

If you build and run this, notice the `text` object will not move if we try to drag it around. `QGraphicsItems` have a `flag` property called `QGraphicsItem::ItemIsMovable` that can be set to allow it to be moved around, either by the user or programmatically:

```
text->setFlag(QGraphicsItem::ItemIsMovable);
```

When we build and run this, you can grab the `text` object and move it around. If it goes beyond our bounding rectangle, it will change text, only returning to the original text if it moves inside the red box again.

If you wanted to animate this, just throw in a timer and change the `text` object's position when the timer fires.

Even with Qt Quick's software renderer, `QGraphicsView` is still a viable solution for graphics animation. If the target device's storage space is really tight, there might not be enough space to add Qt Quick libraries. Or a legacy app might be difficult to import to Qt Quick.

Summary

In this chapter we covered some of the issues facing mobile and embedded developers when trying to develop for smaller display devices, and how `QStyleSheets` can be used to change the interface at runtime to adapt itself for using touchscreen inputs.

We discussed storage and memory space requirements, and the need to configure unneeded features out of Qt to make it have a smaller footprint.

We went through handling orientation changes and discussed using screen gestures such as `Pinch` and `Swipe`.

We learning how to use Qt Designer to add `QLayouts` to create dynamically resizing applications.

Finally, we discussed how to use `QGraphicsView` to utilize graphical elements such as graphical text and images.

Next, we will go through the next best thing since sliced bread for mobile and embedded development—Qt Quick and QML. Then we'll crack on with the real fancy stuff about graphical effects to spice up any interface!

Fluid UI with Qt Quick

My television uses Qt. My phone uses Qt. I could buy a car that uses Qt. I can fly on a plane that uses Qt on its infotainment center. All these things use Qt Quick as their UI. Why? Because it provides faster development—no waiting around for compiling—and the syntax is easy to use, but complex enough to customize it beyond your imagination.

Qt Quick started out being developed in the Brisbane development office of Trolltech as one developer's research project. One of my jobs was to put a demo app of an early version of it onto a Nokia N800 tablet, which I had customized to run Qtopia instead of Nokia's Maemo interface. This was before Nokia purchased the Trolltech company. In my opinion, it was going to become the next generation of Qtopia, which had been renamed Qt Extended. Qtopia, by 2006, had been sold on millions of phone handsets, including 11 models of phones and 30 various handheld devices. Some parts of Qtopia were melded into Qt itself – my favorites, Qt Sensors, and Qt Bearer Management, are examples of these. This new XML-like framework became QML and Qt Quick.

Qt Quick is a really exciting technology and it seems to be taking over the world. It is used in laptops, mobile phones such as the Jolla Sailfish, and medical devices, among others things.

It allows rapid development, fluid transformations, animations, and special effects. Qt Quick allows developers to design customized animated **User Interfaces** (**UI**). Along with the related Qt Quick Controls 2 and Qt Charts APIs, anyone can create snazzy mobile and embedded apps.

In this chapter, we will design and construct an animated UI. We will also cover basic components, such as `Item`, `Rectangle`, and more advanced elements, such as `GraphicsView`. We will look at positioning items with anchors, states, animations, and transitions, and we will also cover traditional features, such as buttons, sliders, and scrollbars. Advanced components showing data in charts, such as BarChart and PieChart, will be shown.

We will be covering the following topics in this chapter:

- Learning Qt Quick basics
- Advanced QML elements in Qt Quick Controls
- Elements for displaying data—Qt Data Visualization and Qt Charts
- Basic animation with Qt Quick

Qt Quick basics – anything goes

Qt Quick is unreal. You should be aware that, at its core, it has only a few fundamental building blocks, called components. You will undoubtedly be using these components quite often:

- `Item`
- `Rectangle`
- `Text`
- `Image`
- `TextInput`
- `MouseArea`

Although there are probably hundreds of components and types, these items are the most important. There are also several classes of elements for text, positioning, states, animation, transitions, and transformations. Views, paths, and data handling all have their own elements.

With those building blocks, you can create fantastic UIs that are alive with animations.

The language to write Qt Quick applications is quite easy to pick up. Let's get started.

QML

Qt Modeling Language (QML) is the declarative programming language that Qt Quick uses. Closely aligned with JavaScript, it is the centerpiece language for Qt Quick. You can use JavaScript functions within a QML document, and Qt Quick will run it.

We use Qt Quick 2 for this book, as Qt Quick 1.0 is depreciated.

All QML documents need to have one or more `import` statements.

This is about the same as C and C++'s `#include` statement.

The most basic QML will have at least one import statement, such as this:

```
import QtQuick 2.12
```

The `.12` corresponds with Qt's minor version, which is the lowest version the application will support.

If you are using properties or components that were added in a certain Qt version, you will need to specify that version.

Qt Quick applications are built with building blocks known as elements, or components. Some basic types are `Rectangle`, `Item`, and `Text`.

Input interaction is supported through `MouseArea` and other items, such as `Flickable`.

One way to start developing a Qt Quick app is by using the Qt Quick app wizard in Qt Creator. You can also grab your favorite text editor and start coding away!

Let's go though some of the following concepts that are important to be aware of as terms that make up the QML language:

- Components, types, and elements
- Dynamic binding
- Signal connection

Components

Components, also known as types or elements, are objects of code and can contain both UI and non-UI aspects.

A UI component example would be the Text object:

```
Text {
// this is a component
}
```

Component properties can be bound to variables, other properties, and values.

Dynamic binding

Dynamic binding is a way to set a property value, which can either be a hardcoded static value, or be bound to other dynamic property values. Here, we bind the Text component's id property to textLabel. We can then refer it to this element just by using its id:

```
Text {
    id: textLabel
}
```

A component can have none, one, or a few signals that can be utilized.

Signal connections

There are two ways signals can be handled. The easiest way is by prepending on and then capitalizing the first letter of the particular signal. For example, a MouseArea has a signal named clicked, which can be connected by declaring onClicked, and then binding this to a function with curly brackets, { }, or even a single line:

```
MouseArea {
    onClicked: console.log("mouse area clicked!")
}
```

You can also use the Connections type to target some other component's signal:

```
Connections {
    target: mouseArea
    onClicked: console.log("mouse area clicked!")
}
```

The model-view paradigm is not dead with Qt Quick. There are a few elements that can show data model views.

Model-view programming

Qt Quick's views are based on a model, which can be defined either with the `model` property or as a list of elements within the component. The view is controlled by a delegate, which is any UI element capable of showing the data.

You can refer to properties of the model data in the delegate.

For example, let's declare a `ListModel`, and fill it with two sets of data. `Component` is a generic object that can be declared, and here, I use it to contain a `Text` component that will function as the delegate. The model's data with the ID of `carModel` can be referred to in the delegate. Here, there is a binding to the `text` property of the `Text` element:

 The source code can be found on the Git repository under the `Chapter02-1b` directory, in the `cp2` branch.

```
ListModel {
    id: myListModel
    ListElement { carModel: "Tesla" }
    ListElement { carModel: "Ford Sync 3" }
}

Component {
    id: theDelegate
    Text {
        text: carModel
    }
}
```

We can then use this model and its delegate in different views. Qt Quick has a few different views to choose from:

- `GridView`
- `ListView`
- `PathView`
- `TreeView`

Let's look at how we can use each of these.

GridView

The GridView type shows model data in a grid, much like a GridLayout.

The grid's layout can be contained with the following properties:

- flow:
    - GridView.FlowLeftToRight
    - GridView.FlowTopToBottom
- layoutDirection:
    - Qt.LeftToRight
    - Qt.RightToLeft
- verticalLayoutDirection:
    - GridView.TopToBottom
    - GridView.BottomToTop

The flow property contains the way the data is presented so it becomes wrapped to the next line or column when it is appropriate. It controls the way it overflows to the next line or column.

The icon for the following example came from https://icons8.com.

FlowLeftToRight means the flow is horizontal. Here's a pictorial representation for FlowLeftToRight:

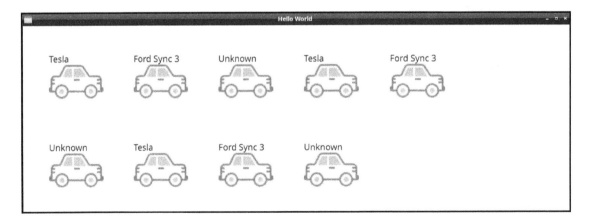

For `FlowTopToBottom`, the flow is vertical; here's a representation of `FlowTopToBottom`:

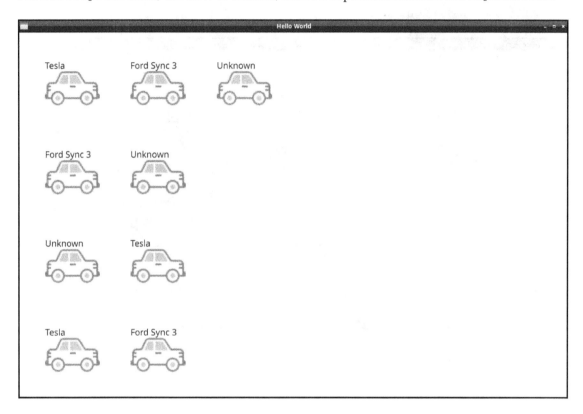

When this example gets built and run, you can resize the window by grabbing the corner with the mouse. You will get a better idea of how the flow works.

The `layoutDirection` property indicates which direction the data will be laid out. In the following case, this is `RightToLeft`:

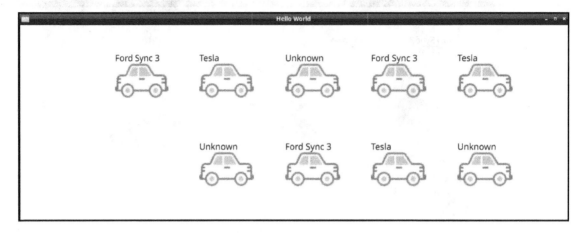

The `verticalLayoutDirection` also indicates which direction the data is laid out, except this will be vertical. Here's the `GridView.BottomToTop` representation:

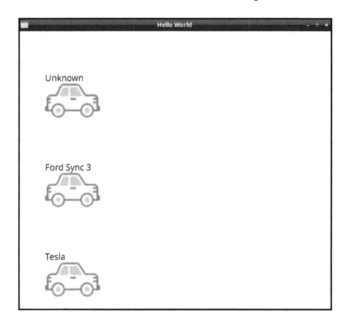

ListView

The QML `Listview` is a type of `Flickable` element, meaning that the user can swipe or flick left or right to progress through the difference views. `ListView` looks different from the `QListView` desktop, as the items are presented in their own page, which is accessible by flicking left or right.

The layout is handled by these properties:

- `orientation:`
  - `Qt.horizontal`
  - `Qt.vertical`
- `layoutDirection:`
  - `Qt.LeftToRight`
  - `Qt.RightToLeft`
- `verticalLayoutDirection:`
  - `ListView.TopToBottom`
  - `ListView.BottonToTop`

PathView

`PathView` shows model data in a `Path`. Its delegate is a view for displaying the model data. It could be a simple drawn line, or an image with text. This can produce a flowing wheel type of data presentation. A `Path` can be constructed by one or more of the following `path` segments:

- `PathAngleArc`: An arc with radii and center
- `PathArc`: An arc with radius
- `PathCurve`: A path through a set of points
- `PathCubic`: A path on Bézier curve
- `PathLine`: A straight line
- `PathQuad`: A quadratic Bézier curve

Here, we use `PathArc` to display a wheel-like item model, using our `carModel`:

 The source code can be found on the Git repository under the `Chapter02-1c` directory, in the `cp2` branch.

```
PathView {
    id: pathView
    anchors.fill: parent
    anchors.margins: 30
    model: myListModel
    delegate:  Rectangle {
        id: theDelegate
        Text {
            text: carModel
        }
        Image {
            source: "/icons8-sedan-64.png"
        }
    }
    path: Path {
        startX: 0; startY: 40
        PathArc { x: 0; y: 400; radiusX:5; radiusY: 5 }
    }
}
```

You should now see something like this:

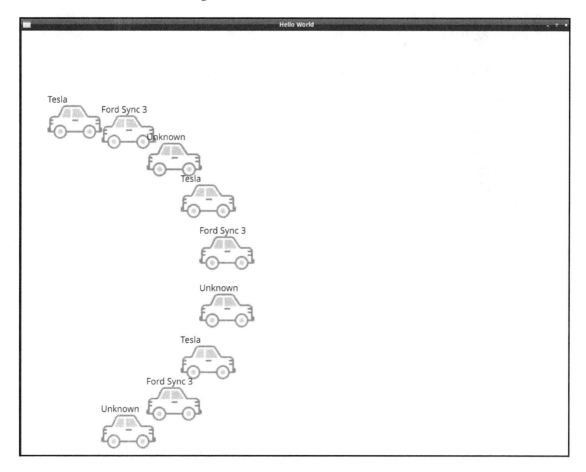

There are a couple of special `path` segments that augment and change attributes of the `path`:

- `PathAttribute`: Allows an attribute to be specified at certain points along a path
- `PathMove`: Moves a path to a new position

TreeView

TreeView is perhaps the most recognizable of these views. It looks very similar to the desktop variety. It displays a tree structure of its model data. TreeView has headers, called TableViewColumn, which you can use to add a title as well as to specify its width. Further customization can be made using headerDelegate, itemDelegate, and rowDelegate.

Sorting is not implemented by default, but can be controlled by a few properties:

- sortIndicatorColumn: Int, indicating the column to be sorted
- sortIndicatorVisible: Bool is used to enable sorting
- sortIndicatorOrder: Enum either Qt.AscendingOrder or Qt.DescendingOrder

Gestures and touch

Touch gestures can be an innovative way to interact with your application. To use the QtGesture class in Qt, you will need to implement the handlers in C++ by overriding the QGestureEvent class and handling the built-in Qt::GestureType. In this way, the following gestures can be handled:

- Qt::TapGesture
- Qt::TapAndHoldGesture
- Qt::PanGesture
- Qt::PinchGesture
- Qt::SwipeGesture
- Qt::CustomGesture

The Qt::CustomGesture flag is a special one that can be used to invent your own custom gestures.

There is one built-in gesture item in Qt Quick— PinchArea.

PinchArea

PinchArea handles pinch gestures, which are commonly used on mobile phones to zoom in on an image from within Qt Quick, so you can use simple QML to implement it for any Item-based element.

You can use the `onPinchFinished`, `onPinchStarted`, and `onPinchUpdated` signals, or set the `pinch.target` property to the target item to handle the pinch gesture.

MultiPointTouchArea

The `MultiPointTouchArea` is not a gesture, but rather a way to track multiple points of contact of the touchscreen. Not all touchscreens support multi-touch. Mobile phones usually support multi-touch, and some embedded devices do as well.

To use multi-point touchscreens in QML, there is the `MultiPointTouchArea` component, which works a bit like `MouseArea`. It can operate alongside `MouseArea` by setting its `mouseEnabled` property to `true`. This makes the `MultiPointTouchArea` component ignore events from the mouse and only respond to touch events.

Each `MultiPointTouchArea` takes an array of `TouchPoints`. Note the use of square brackets, [ ]—this denotes that it is an array. You can define one or more of these to handle a certain number of `TouchPoints` or fingers. Here, we define and handle only three `TouchPoints`.

If you try this on a non-touchscreen, only one green dot will track the touch point:

 The source code can be found on the Git repository under the `Chapter02-2a` directory, in the `cp2` branch.

```
import QtQuick 2.12
import QtQuick.Window 2.12
Window {
    visible: true
    width: 640
    height: 480
    color: "black"
    title: "You can touch this!"

    MultiPointTouchArea {
        anchors.fill: parent
        touchPoints: [
            TouchPoint { id: touch1 },
            TouchPoint { id: touch2 },
            TouchPoint { id: touch3 }
        ]
        Rectangle {
            width: 45; height: 45
            color: "#80c342"
```

```
                    x: touch1.x
                    y: touch1.y
                    radius: 50
                    Behavior on x  {
                        PropertyAnimation {easing.type: Easing.OutBounce;
duration: 500 }
                    }
                    Behavior on y  {
                        PropertyAnimation {easing.type: Easing.OutBounce;
duration: 500 }
                    }
            }
        Rectangle {
            width: 45; height: 45
            color: "#b40000"
            x: touch2.x
            y: touch2.y
            radius: 50
                Behavior on x  {
                    PropertyAnimation {easing.type: Easing.OutBounce;
duration: 500 }
                    }
                Behavior on y  {
                    PropertyAnimation {easing.type: Easing.OutBounce;
duration: 500 }
                    }
        }
        Rectangle {
            width: 45; height: 45
            color: "#6b11d8"
            x: touch2.x
            y: touch2.y
            radius: 50
                Behavior on x  {
                    PropertyAnimation {easing.type: Easing.OutBounce;
duration: 500 }
                    }
                Behavior on y  {
                    PropertyAnimation {easing.type: Easing.OutBounce;
duration: 500 }
                    }
            }
        }
}
```

You should see this when you run it on a non-touchscreen:

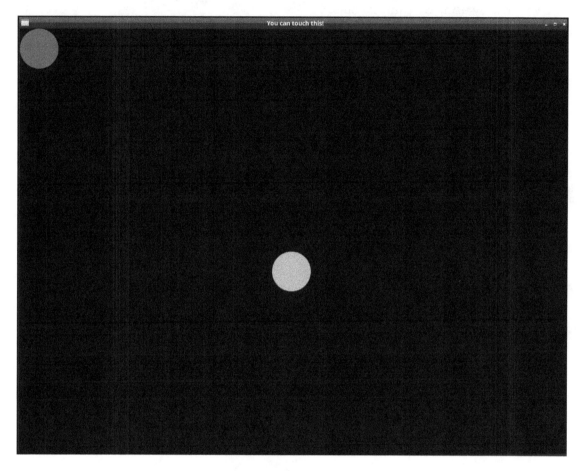

Notice the `PropertyAnimation`? We'll get to that soon; keep reading.

Positioning

With the myriad of different mobile phones and embedded device sizes currently available, the dynamic positioning of elements becomes more important. You may not necessarily want things placed randomly on the screen. If you have a great-looking layout on a high-DPI iPhone, it may look completely different on a small Android device, with images covering half of the screen. Automatic layouts in QML are called positioners.

Mobile and embedded devices come with a variety of screen sizes. We can better target the size variance by using dynamic layouts.

Layouts

These are the positioning elements that arrange the layout of the different items that you may want to use:

- `Grid`: Positions items in a grid
- `Column`: Positions items vertically
- `Row`: Positions items horizontally
- `Flow`: Positions items side by side with wrapping

Additionally, there are also the following items:

- `GridLayout`
- `ColumnLayout`
- `RowLayout`
- `StackLayout`

The difference between the `Grid` and the `GridLayout` elements are that the layouts are more dynamic in terms of resizing. Layouts have attached properties, so you can easily specify aspects of the layout, such as `minimumWidth`, the number of columns, or the number of rows. The item can be made to fill itself to the grid or fixed width.

You can also use *rigid* layouts which are more like tables. Let's look at using layouts that are slight less dynamic and use static sizing.

Rigid layouts

I use the word *rigid* because they are less dynamic than all the layout items. The cell sizes are fixed and based on a percentage of the space where they are contained. They cannot span across rows or columns to fill the next column or row. Take this code, for example.

It has no layouts at all, and, when you run it, all the elements get squished together on top of one another:

 The source code can be found on the Git repository under the `Chapter02-3` directory, in the `cp2` branch.

```
import QtQuick 2.12
import QtQuick.Window 2.12

Window {
    visible: true
    width: 640
    height: 480
    title: qsTr("Hello World")
    Rectangle {
        width: 35
        height: 35
        gradient: Gradient {
            GradientStop { position: 0.0; color: "green"; }
            GradientStop { position: 0.25; color: "purple"; }
            GradientStop { position: 0.5; color: "yellow"; }
            GradientStop { position: 1.0; color: "black"; }
        }
    }
    Text {
        text: "Hands-On"
        color: "purple"
        font.pointSize: 20
    }
    Text {
        text: "Mobile"
        color: "red"
        font.pointSize: 20
    }
    Text {
        text: "and Embedded"
        color: "blue"
        font.pointSize: 20
    }
}
```

As you can see in the following screenshot, all of the elements are bunched up on top of each other without positioning:

This was probably not what the design team had dreamed up. Unless, of course, they did, and then wanted to use a `PropertyAnimation` value to animate the elements moving to their proper layout positions.

What happens when we add a `Column` QML element? Examine the following code:

The source code can be found on the Git repository under the `Chapter02-3a` directory, in the `cp2` branch.

```
Rectangle {
  width: 500
  height: 500
      Column {
          Rectangle {
              width: 35
              height: 35
              gradient: Gradient {
```

```
                GradientStop { position: 0.0; color: "green"; }
                GradientStop { position: 0.25; color: "purple"; }
                GradientStop { position: 0.5; color: "yellow"; }
                GradientStop { position: 1.0; color: "black"; }
            }
        }

        Text {
            text: "Hands-On"
            color: "purple"
            font.pointSize: 20
        }

        Text {
            text: "Mobile"
            color: "red"
            font.pointSize: 20
        }

        Text {
            text: "and Embedded"
            color: "blue"
            font.pointSize: 20
        }
    }
}
```

When you build this example, the layout looks like this:

That's more like what the designer's mock-ups look like! (I know; cheap designers.)

`Flow` is another layout item we can use.

The source code can be found on the Git repository under the Chapter02-3b directory, in the cp2 branch.

```
Flow {
    anchors.fill: parent
    anchors.margins: 4
    spacing: 10
```

Now, from our preceding code, change Column to Flow, add some anchor items, and build, then run on a simulator to get a feel for how the Flow item works on a small screen:

The Flow type will wrap its contents around if needed, and, indeed, it has wrapped here on the last Text element. If this were to be re-oriented to the landscape orientation or on a tablet, there would be no need to wrap, and all of these elements would be on one row at the top.

Dynamic layout

Instead of using a `Grid` element to lay out items, there is also `GridLayout`, which can be used to customize the layout. In terms of targeting mobile and embedded devices that come with different screen sizes and device orientations, it is probably better to use `GridLayout`, `RowLayout`, and `ColumnLayout`. Using these, you will gain the ability to use its attached properties. Here is a list of attached properties you can use:

`Layout.alignment`	A `Qt.Alignment` value specifying alignment of item within the cell
`Layout.bottomMargin`	Bottom margin of space
`Layout.column`	Specifies column position
`Layout.columnSpan`	How many columns to spreads out to
`Layout.fillHeight`	If `true`, item fills to the height
`Layout.fillWidth`	If `true`, item fills to the width
`Layout.leftMargin`	Left margin of space
`Layout.margins`	All margins of space
`Layout.maximumHeight`	Maximum height of item
`Layout.maximumWidth`	Maximum width of item
`Layout.minimumHeight`	Minimum height of item
`Layout.minimumWidth`	Minimum width of item
`Layout.preferredHeight`	Preferred height of item
`Layout.preferredWidth`	Preferred width of item
`Layout.rightMargin`	Right margin of space
`Layout.row`	Specifies row position
`Layout.rowSpan`	How many rows to spread out to
`Layout.topMargin`	Top margin of space

In this code, we use `GridLayout` to position the three `Text` items. The first `Text` item will span, or fill, two rows so that the second `Text` will be in the second row:

 The source code can be found on the Git repository under the `Chapter02-3c` directory, in the `cp2` branch.

```
GridLayout {
    rows: 3
    columns: 2
    Text {
        text: "Hands-On"
```

```
        color: "purple"
        font.pointSize: 20
    }
    Text {
        text: "Mobile"
        color: "red"
        font.pointSize: 20
    }
     Text {
        text: "and Embedded"
        color: "blue"
        font.pointSize: 20
        Layout.fillHeight: true
    }
}
```

Positioning is a way to get dynamically changing applications and allow them to work on various devices without having to change the code. `GridLayout` works much like a layout, but with expanded capabilities.

Let's take a look at how we can dynamically position these components using `Anchors`.

Anchors

`Anchors` are related to positioning, and are a way to position elements relative to each other. They are a way to dynamically position UI elements and layouts.

They use the following points of contact:

- `left`
- `right`
- `top`
- `bottom`
- `horizontalCenter`
- `verticalCenter`

Take, for example, two images; you can put them together like a puzzle by specifying anchor positions:

```
Image{ id: image1; source: "image1.png"; }
Image{ id: image2; source: "image2.png; anchors.left: image1.right; }
```

This will position the left side of image2 at the right side of image1. If you were to add anchors.top: parent.top to image1, both of these items would then be positioned relative to the top of the parent position. If the parent was a top-level item, they would be placed at the top of the screen.

Anchors are a way to achieve columns, rows, and grids of components that are relative to some other component. You can anchor items diagonally and anchor them apart from each other, among other things.

For example, the anchor property of Rectangle, called fill, is a special term meaning top, bottom, left, and right, and is bound to its parent. This means that it will fill itself to the size of its parent.

Using anchors.top indicates an anchor point for the top of the element, meaning that it will be bound to the parent component's top position. For example, a Text component will sit above of the Rectangle component.

To get a component such as Text to be centered horizontally, we use the anchor.horizontal property and bind it with the parent.horizontalCenter positional property.

Here, we anchor the Text label to the top center of the Rectangle label, itself anchored to fill its parent, which is the Window:

```
import QtQuick 2.12
import QtQuick.Window 2.12

Window {
    visible: true
    width: 500
    height: 500

    Rectangle {
      anchors.fill: parent
        Text {
            id: textLabel
            text: "Hands-On Mobile and Embedded"
            color: "purple"
            font.pointSize: 20
            anchors.top: parent.top
            anchors.horizontalCenter: parent.horizontalCenter
        }
    }
}
```

The source code can be found on the Git repository under the `Chapter02` directory, in the `cp2` branch.

The `Window` component was provided by the Qt Quick app wizard and is not visible by default, so the wizard set the `visible` property to `true` as we need see it. We will use `Window` as the parent for the `Rectangle` component. Our `Rectangle` component will provide an area for our `Text` component, which is a simple label type.

Each component has its own properties to fiddle with. By fiddling, I mean binding. For instance, the `color:` `"purple"` line is binding the color referenced as "purple" to the `color` property of the `Text` element. These bindings do not have to be static; they can be dynamically changed, and the property's value that they are bound to changes as well. This value binding will persist until the property is written with another value.

The background of this application is boring. How about we add a gradient there? Under the closing bracket for the `Text` component, but still within the `Rectangle`, add this gradient. `GradientStop` is a way to specify a color at a certain point in the gradient. The `position` property is a percent fraction point from zero to one, corresponding to where the color should start. The gradient will fill in the gap in between:

```
gradient: Gradient {
    GradientStop { position: 0.0; color: "green"; }
    GradientStop { position: 0.25; color: "purple"; }
    GradientStop { position: 0.75; color: "yellow"; }
    GradientStop { position: 1.0; color: "black"; }
}
```

The source code can be found on the Git repository under the `Chapter02-1` directory, in the `cp2` branch.

As you can see, the gradient starts with the green color at the top, smoothly blends to purple, then yellow, and finishes at black:

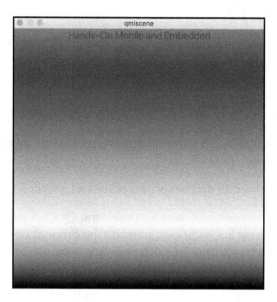

Easy peasy, lemon squeezy!

Layouts and anchors are important to be able to control the UIs. They provide an easy way to deal with differences in display size and orientation changes on hundreds of different devices with different screen sizes. You could have a QML file work on all displays, although it is recommended to use different layouts for extremely different devices. An application could work fine on a tablet, or even a phone, but try to place it on a watch or other embedded device, and you will run into trouble accessing many details that your users can use.

Qt Quick has many building blocks to create useful applications on any device. What happens when you don't want to create all the UI elements yourself? That is where Qt Quick Controls come into play.

Qt Quick Controls 2 button, button, who's got the button?

For a while in the life of Qt Quick, there were only basic components, such as `Rectangle` and `Text`. Developers had to create their own implementations of buttons, dials, and just about every common UI item. As it matured, it also grew elements such as `Window` and even `Sensor` elements. There were always rumblings about having a set of common UI elements available. Eventually, common UI elements were released.

Spotlight on Qt Quick Controls. No more having to create your own buttons and other components, yay! And developers rejoiced!

Then, they found a better way to do things and released Qt Quick Controls 2!

 Qt Quick Controls comes in two versions, Qt Quick Controls and Qt Quick Controls 2. Qt Quick Controls (the original one) has been depreciated by Qt Quick Controls 2. Any new use of these components should use Qt Quick Controls 2.

You can access all kinds of common UI elements, including the following:

- `Buttons`
- `Containers`
- `Input`
- `Menu`
- `Radio buttons`
- `Progress bar`
- `Popups`

Let's examine a simple Qt Quick Controls 2 example.

An `ApplicationWindow` has attached `menuBar`, `header`, and `footer` properties that you can use to add whatever you need to them. Since an `ApplicationWindow` is not visible by default, we almost always need to add `visible: true`.

Here, we will add a traditional menu with `TextField` in the header.

The menu has an `onTriggered` signal, which is used here to run the `open()` function of `MessageDialog`:

 The source code can be found on the Git repository under the `Chapter02-4` directory, in the cp2 branch.

```
import QtQuick 2.12
import QtQuick.Controls 2.3
import QtQuick.Dialogs 1.1

ApplicationWindow {
    visible: true
    title: "Mobile and Embedded"
    menuBar: MenuBar {
        Menu { title: "File"
            MenuItem { text: "Open "
                onTriggered: helloDialog.open()
            }
        }
    }
    header: TextField {
        placeholderText: "Remember the Qt 4 Dance video?"
    }
    MessageDialog {
        id: helloDialog
        title: "Hello Mobile!"
        text: "Qt for Embedded devices to rule the world!"
    }
}
```

Here's what our code would result in:

Oooooh – fancy!

Qt Quick Controls 2 has different styles to choose from – `Default`, `Fusion`, `Imagine`, `Material`, and `Universal`. This can be set in the C++ backend as `QQuickStyle::setStyle("Fusion");`. I presume you do have a C++ backend, right?

Views that can come in handy on mobile and embedded devices are as follows:

- `ScrollView`
- `StackView`
- `SwipeView`

These can be helpful on small screens, as they provide a way to easily view and access several pages without too much hassle. A `Drawer` element is also handy and can provide a way to implement a menu or a toolbar that sticks to the side.

Buttons are awesome, and Qt Quick Controls 2 has buttons. It even has the `RoundButton` component, as well as icons for the buttons! Before Qt Quick Controls, we had to roll these up ourselves. At the same time, it is nice that we can implement these things to do what we need with little effort. And now with even less effort!

Let's put some of these to the test and expand upon our last example.

I like `SwipeView`, so let's use that, with two `Page` elements as children of `SwipeView`:

 The source code can be found on the Git repository under the `Chapter02-5` directory, in the `cp2` branch.

```
SwipeView {
    id: swipeView
    anchors.fill: parent
    Page {
        id: page1
        anchors.fill: parent.fill
        header: Label {
            text: "Working"
            font.pixelSize: Qt.application.font.pixelSize * 2
            padding: 10
        }
        BusyIndicator {
            id: busyId
            anchors.centerIn: parent
            running: true;
        }
        Label {
```

```
                    text: "Busy Working"
                    anchors.top: busyId.bottom
                    anchors.horizontalCenter: parent.horizontalCenter
                }
            }

        Page {
            id: page2
            anchors.fill: parent.fill
            header: Label {
                text: "Go Back"
                font.pixelSize: Qt.application.font.pixelSize * 2
                padding: 10
            }
            Label {
                text: "Nothing here to see. Move along, move along."
                anchors.centerIn: parent
            }
        }
    }

    PageIndicator {
        id: indicator
        count: swipeView.count
        currentIndex: swipeView.currentIndex
        anchors.bottom: swipeView.bottom
        anchors.horizontalCenter: parent.horizontalCenter
    }
```

I think that a `PageIndicator` at the bottom to indicate which page we are on gives the user some visual feedback for navigation. We tie in `PageIndicator` by binding the `count` of `SwipeView` and `currentIndex` properties to its properties of the same name. How convenient!

Instead of `PageIndicator`, we could just as easily use `TabBar`.

Customizing

You can customize the look and feel of just about every Qt Quick Control 2 component. You can override different properties of the controls, such as background. In the previous example code, we customized the Page header. Here, we override the background to a button, add our own Rectangle, color it, give it a border with a contrasting color, and make it rounded at the ends by using the radius property. Here's how it would work:

The source code can be found on the Git repository under the Chapter02-5 directory, in the cp2 branch.

```
Button {
    text: "Click to go back"
    background: Rectangle {
        color: "#673AB7"
        radius: 50
        border.color: "#4CAF50"
        border.width: 2
    }
    onClicked: swipeView.currentIndex = 0
}
```

Customizing is easy with Qt Quick. It was built with customizing in mind. The ways are endless. Nearly all the Qt Quick Controls 2 elements have visual elements that can be customized including most of the background and content items, although not all.

These controls seem to be best on a desktop, but they can be customized to work well on mobile and embedded devices. The ScrollBar property of ScrollView can be made larger in width on touchscreens.

Show your data – Qt Data Visualization and Qt Charts

Qt Quick has a convenient way to show data of all kinds. The two modules, Qt Data Visualization and Qt Charts, can both supply integral UI elements. They are similar, except Qt Data Visualization displays data in 3D.

Qt Charts

Qt Charts shows 2D graphs and uses the Graphics View framework.

It adds the following chart types:

- Area
- Bar
- Box-and-whiskers
- Candlestick
- Line: a simple line chart
- Pie: pie slices
- Polar: a circular line
- Scatter: a collection of points
- Spline: a line chart with curved points

The following example from Qt shows a few different charts that are available:

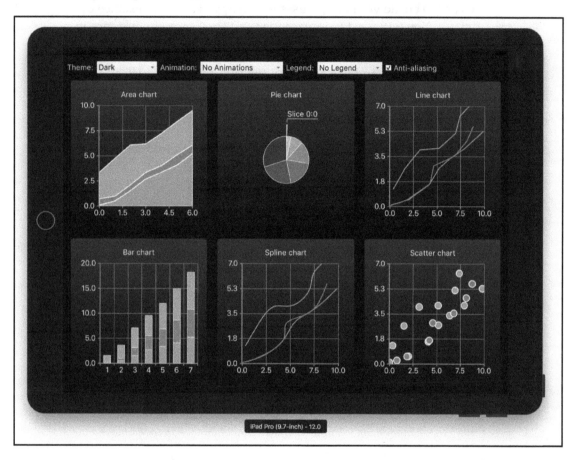

Each graph or chart has at least one axis and can have the following types:

- Bar axis
- Category
- Date-time
- Logarithmic value
- Value

Qt Charts requires a `QApplication` instance. If you use the Qt Creator wizard to create your app, it uses a `QGuiApplication` instance by default. You will need to replace the `QGuiApplication` instance in `main.cpp` to `QApplication`, and also change the `includes` file.

You can use grid lines, shades, and tick marks on the axis, which can also be shown in these charts.

Let's look at how to create a simple BarChart.

 The source code can be found on the Git repository under the `Chapter02-6` directory, in the `cp2` branch.

```
import QtCharts 2.0
ChartView {
    title: "Australian Rain"
    anchors.fill: parent
    legend.alignment: Qt.AlignBottom
    antialiasing: true

    BarSeries {
        id: mySeries
        axisX: BarCategoryAxis {
            categories: ["2015", "2016", "2017" ]
        }
        BarSet { label: "Adelaide"; values: [536, 821, 395] }
        BarSet { label: "Brisbane"; values: [1076, 759, 1263] }
        BarSet { label: "Darwin"; values: [2201, 1363, 1744] }
        BarSet { label: "Melbourne"; values: [526, 601, 401] }
        BarSet { label: "Perth"; values: [729, 674, 578] }
        BarSet { label: "Sydney"; values: [1076, 1386, 1338] }
    }
}
```

See how nice the charts look? Have a look:

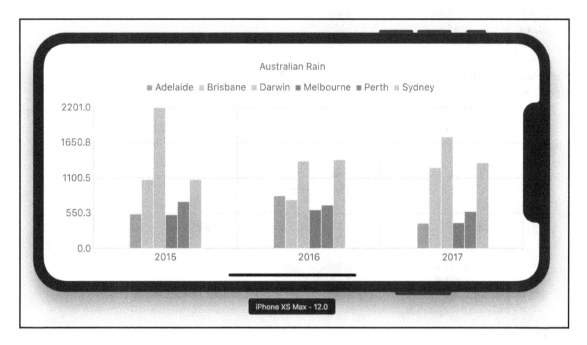

Qt Data Visualization

Qt Data Visualization is similar to Qt Charts but presents data in 3D form. It can be downloaded through Qt Creator's Maintenance Tool app. It is available for use with Qt Widget and Qt Quick. We will be working with the Qt Quick version. It uses OpenGL to present 3D graphs of data.

Since we are targeting mobile phones and embedded devices, we talk about using OpenGL ES2. There are some features of Qt Data Visualization that do not work with OpenGl ES2, which is what you will find on mobile phones:

- Antialiasing
- Flat shading
- Shadows
- Volumetric objects that use 3D textures

Let's try using a `Bars3D` with data from the total amount of rain in certain Australian cities used in the previous example.

I set the theme to `Theme3D.ThemeQt`, which is a green-based theme. Add a few customizations such as font size to be able to see the content better on small mobile displays.

`Bar3DSeries` will manage the visual elements such as labels for rows, columns, and the data, which here is the total rain amount for that year. `ItemModelBarDataProxy` is the proxy for displaying the data. The model data here is a `ListModel` containing `ListElement`s of cities rainfall data for the previous three years. We will use the same data from the previous Qt Charts example so you can compare the differences in the way the bar charts display their data:

 The source code can be found on the Git repository under the `Chapter02-7` directory, in the `cp2` branch.

```
import QtQuick 2.12
import QtQuick.Window 2.12
import QtDataVisualization 1.2
Window {
    visible: true
    width: 640
    height: 480
    title: qsTr("Australian Rain")
    Bars3D {
        width: parent.width
        height: parent.height
        theme: Theme3D {
            type: Theme3D.ThemeQt
            labelBorderEnabled: true
            font.pointSize: 75
            labelBackgroundEnabled: true
        }
        Bar3DSeries {
            itemLabelFormat: "@colLabel, @rowLabel: @valueLabel"
            ItemModelBarDataProxy {
                itemModel: dataModel
                rowRole: "year"
                columnRole: "city"
                valueRole: "total"
            }
        }
    }
    ListModel {
        id: dataModel
        ListElement{ year: "2017"; city: "Adelaide"; total: "536"; }
```

```
        ListElement{ year: "2016"; city: "Adelaide"; total: "821"; }
        ListElement{ year: "2015"; city: "Adelaide"; total: "395"; }
        ListElement{ year: "2017"; city: "Brisbane"; total: "1076"; }
        ListElement{ year: "2016"; city: "Brisbane"; total: "759"; }
        ListElement{ year: "2015"; city: "Brisbane"; total: "1263"; }
        ListElement{ year: "2017"; city: "Darwin"; total: "2201"; }
        ListElement{ year: "2016"; city: "Darwin"; total: "1363"; }
        ListElement{ year: "2015"; city: "Darwin"; total: "1744"; }
        ListElement{ year: "2017"; city: "Melbourne"; total: "526"; }
        ListElement{ year: "2016"; city: "Melbourne"; total: "601"; }
        ListElement{ year: "2015"; city: "Melbourne"; total: "401"; }
        ListElement{ year: "2017"; city: "Perth"; total: "729"; }
        ListElement{ year: "2016"; city: "Perth"; total: "674"; }
        ListElement{ year: "2015"; city: "Perth"; total: "578"; }
        ListElement{ year: "2017"; city: "Sydney"; total: "1076"; }
        ListElement{ year: "2016"; city: "Sydney"; total: "1386"; }
        ListElement{ year: "2015"; city: "Sydney"; total: "1338"; }
    }
}
```

You can run this on a touchscreen device, and then move the chart around in 3D!:

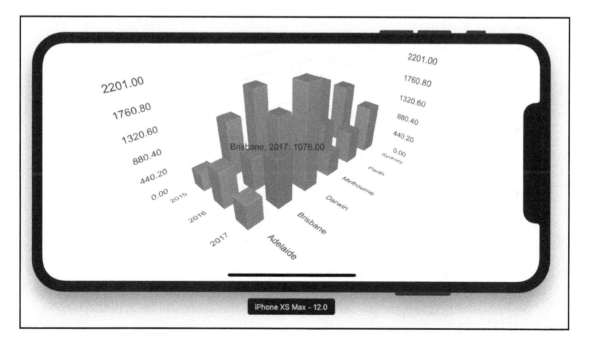

You can grab the graph and spin it around to see the data from different perspectives. You can zoom in and back out, as well.

The `QtDataVisualization` module also has scatter and surface graphs that show data in 3D.

Animate it!

This is where it gets gloriously complicated. There are various types of animations:

- `ParallelAnimation`
- `SmoothedAnimation`
- `PauseAnimation`
- `SequentialAnimation`

Additionally, `PropertyAction` and `ScriptAction` can be used. A `PropertyAction` is a change to any property that happens without an animation. We learned about `ScriptAction` in the last section on *States*.

There are also animation types that operate on various values:

- `AnchorAnimation`
- `ColorAnimation`
- `NumberAnimation`
- `OpacityAnimator`
- `PathAnimation`
- `ParentAnimation`
- `PropertyAnimation`
- `RotationAnimation`
- `SpringAnimation`
- `Vector3DAnimation`

A `Behavior` can be used to specify an animation for a property change.

Let's look at how some of these can be used.

Transitions

Transitions and states are explicitly tied together. A `Transition` animation happens when a `State` change occurs.

A `State` change can handle different kinds of changes:

- `AnchorChanges`: Changes to anchor layouts
- `ParentChanges`: Changes to parenting (as in reparenting)
- `PropertyChanges`: Changes to the target's properties

You can even run JavaScript on `State` changes using `StateChangeScript` and `ScriptAction`.

To define different `states`, an element has a `states` array of `State` elements that can be defined. We will add a `PropertyChanges`:

```
states : [
    State {
        name: "phase1"
        PropertyChanges { target: someTarget; someproperty: "some value"; }
    },
    State {
        name: "phase2"
        PropertyChanges { target: someTarget; someproperty: "some other
value"; }
    }
]
```

Target properties can be just about anything—`opacity`, `position`, `color`, `width`, or `height`. If an element has a changeable property, the chances are that you can animate it in a `State` change.

As I mentioned before, to run a script in a `State` change, you can define a `StateChangeScript` in the `State` element that you want it to run in. Here, we simply output some logging text:

```
function phase3Script() {
    console.log("demonstrate a state running a script");
}

State {
    name: "phase3"
    StateChangeScript {
        name: "phase3Action"
        script: phase3Script()
    }
}
```

Just imagine the possibilities! We haven't even presented animations! We will go there next.

Animation

Animation can spice up your apps in wonderful ways. Qt Quick makes it almost trivial to animate different aspects of your application. At the same time, it allows you to customize them into unique and more complicated animations.

PropertyAnimation

`PropertyAnimation` animates an item's changeable property. Typically, this is x or y color, or it can be some other property of any item:

```
Behavior on activeFocus { PropertyAnimation { target: myItem; property:
color; to: "green"; } }
```

The `Behavior` specifier implies that when the `activeFocus` is on `myItem`, the `color` will change to `green`.

NumberAnimation

`NumberAnimation` derives from `PropertyAnimation`, but only works on properties that have a `qreal` changeable value:

```
NumberAnimation { target: myOtherItem; property: "y"; to: 65; duration: 250
}
```

This will move the `myOtherItem` element's y position to 65 over a 250-microsecond period of time.

Some of these animation elements control how other animations are played, including `SequentialAnimation` and `ParallelAnimation`.

SequentialAnimation

`SequentialAnimation` is an animation that runs other animation types consecutively, one after the other, like a numbered procedure:

```
SequentialAnimation {
    NumberAnimation { target: myOtherItem; property: "x"; to: 35; duration:
1500 }
    NumberAnimation { target: myOtherItem; property: "y"; to: 65; duration:
1500 }
}
```

In this instance, the animation that would play first is `ColorAnimation`, and, once that is finished, it would play `NumberAnimation`. Move the `myOtherItem` element's x property to position 35, and then move its y property to position 65, in two steps:

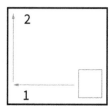

You can use either on `<property>` or `properties` to target a property.

Also available is the `when` keyword, which denotes when something can take place. It can be used with any property if it evaluates to `true` or `false`, such as `when: y > 50`. You could use it, for example, on the `running` property.

ParallelAnimation

`ParallelAnimation` plays all its defined animations at the same time, asynchronously:

```
ParallelAnimation {
    NumberAnimation { target: myOtherItem; property: "x"; to: 35; duration:
1500 }
    NumberAnimation { target: myOtherItem; property: "y"; to: 65; duration:
1500 }
}
```

These are the same animations, but this would perform them at the same time.

It is interesting to note that this animation would move `myOtherItem` to position 35 and 65 directly from where the current position is, as if it were one step:

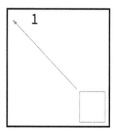

SpringAnimation

`SpringAnimation` animates items with a spring-like motion. It has two properties to pay attention to—`spring` and `damping`:

- `spring`: A `qreal` value that controls how energetic the bounce is
- `damping`: How quickly the bouncing stops
- `mass`: Adds a weight to the bounce, so it acts as if there is gravity and weight
- `velocity`: Specifies the maximum velocity
- `modulus`: The value at which a value will wrap around to zero
- `epsilon`: Amount of rounding to zero

 The source code can be found on the Git repository under the `Chapter02-8` directory, in the `cp2` branch.

```
import QtQuick 2.12
import QtQuick.Window 2.12
Window {
    visible: true
    width: 640
    height: 480
    color: "black"
    title: qsTr("Red Bouncy Box")
    Rectangle {
        id: redBox
        width: 50; height: 50
        color: "black"
        border.width: 4
        border.color: "red"
        Behavior on x { SpringAnimation { spring: 10; damping: 10; } }
        Behavior on y { SpringAnimation { spring: 10; damping: .1;  mass:
10 } }
    }
    MouseArea {
        anchors.fill: parent
        hoverEnabled: true
        onClicked: animation.start()
        onPositionChanged: {
            redBox.x = mouse.x - redBox.width/2
            redBox.y = mouse.y - redBox.height/2
        }
    }
    ParallelAnimation {
```

```
        id: animation
        NumberAnimation { target: redBox; property: "x"; to: 35; duration:
  1500 }
        NumberAnimation { target: redBox; property: "y"; to: 65; duration:
  1500 }
      }
  }
```

In this example, a red square follows the finger or mouse cursor around, bouncing up and down as it goes. When the user clicks on the app, the red square will move to position 35 and 65. A `spring` value of 10 makes it very bouncy, but the `mass` of 10 on the y axis will cause it to bounce like it has more weight. The lower the `damping` value is, the more quickly it will come to rest. Here, the `damping` value is much greater on the x axis, so it will tend to keep bouncing up and down more than side to side.

Easing

I should mention easing at this point. Every Qt Quick animation has an `easing` property. Easing is a way to specify the speed at which the animation progresses. The default `easing` value is `Easing.Linear`. There are 40 different `easing` properties, which are probably better seen running in an example than seen here demonstrated with graphs.

You can see a demonstration of this at my GitHub web server by the magic of Qt for WebAssembly at

`https://lpotter.github.io/easing/easing.html`.

Qt for WebAssembly brings Qt apps to the web. Firefox has the fastest WebAssembly implementation at the time of writing this book. We will discuss Qt for WebAssembly in `Chapter 14`, *Universal Platform for Mobiles and Embedded Devices*.

SceneGraph

Scene Graph is based on OpenGL for Qt Quick. On mobile and embedded devices, it is usually OpenGL ES2. As I mentioned before, Scene Graph caters to manage a sizable number of graphics. OpenGL is a huge subject worthy of its own book—in fact, tons of books—about OpenGL ES2 programming. I won't go into too much detail about it here, but will just mention that OpenGL is available for mobile phones and embedded devices, depending on the hardware.

If you are planning to use Scene Graph, most of the heavy lifting will be done in C++. You should already be familiar with how to use C++ and QML together, as well as OpenGL ES2. If not, Qt has great documentation on it.

Summary

Qt Quick is ready-made for using on mobile and embedded devices. From the simple building blocks of basic Qt Quick items to 3D data charts, you can write complicated animated applications using various data sets and presentations in QML.

You should now be able to use basic components such as `Rectangle` or `Text` to create Qt Quick applications that use dynamic variable bindings and signals.

We also covered how to use `anchors` to position the components visually and will be able to accept changing orientations and various screen sizes of target devices.

You are now able to use more conventional-looking components such as ready-made `Button`, `Menu` and `ProgressBar` instances, as well as more advanced graphical elements such as `PieChart` and `BarChart`.

We also examined using different animation methods available in Qt Quick, such as `ProperyAnimation` and `NumberAnimation`.

In the next chapter, we will learn about using particles and special graphical effects.

3
Graphical and Special Effects

Qt Quick has been extended with animation and special effects through the use of particles. Particles and Qt Graphical Effects will make a **User Interface (UI)** come alive and stand out among the crowd.

The particle system in Qt Quick allows for a large number of images or other graphical objects to simulate highly energized and chaotic animation and effects. Simulating snow falling or explosions with fire is made easier by using a particle system. Dynamic properties of these elements animate these even more.

Using Qt Graphical Effects can help make UIs visually more appealing and make it easier for the user to differentiate between graphical components. Drop shadows, glows, and blurring make 2-dimensional objects seem more like 3-dimensional ones.

In this chapter, we will cover the following topics:

- The universe of particles
- Particle `painters`, `emitters`, and `affectors`
- Graphical effects for Qt Quick

The universe of particles

Finally! We have reached the fun part of the book where the magic happens. It's all fine and dandy using rectangles, text, and buttons, but particles add splash and zing, together with adding wisps of light to games. They can also be used to highlight and emphasize items of interest.

Particles are a type of animation that consist of numerous graphical elements, all moving in a fuzzy manner. There are four main QML components to use:

- `ParticleSystem`: Maintains the particle animation timeline
- `Emitters`: Radiates the particles in to the system

- `Painters`: These components paint the particles. Here are the various components:
  - `ImageParticle`: A particle using an image
  - `ItemParticle`: A particle using a QML item as delegate
  - `CustomParticle`: A particle using a shader
- `Affectors`: Alters the properties of a particle

To see how we manage all these items, let's take a look at the main particle manager, the `ParticleSystem`.

ParticleSystem

The `ParticleSystem` component maintains the particle animation timeline. It is what bonds all the other elements together and acts as the center for operations. You can `pause`, `resume`, `restart`, `reset`, `start`, and `stop` the particle animation.

The `painters`, `emitters`, and `affectors` all interact with each other through the `ParticleSystem`.

Many `ParticleSystem` components can exist in your application, and each has an `Emitter` component.

Let's dive a little more into the details about particle `painters`, `emitters`, and `affectors`.

Particle painters, emitters, and affectors

Particles in Qt Quick are graphical elements such as images, QML items, and OpenGL shaders.

They can be made to move and flow in endless ways.

Every particle is part of a `ParticleGroup`, which, by default, has an empty name. A `ParticleGroup` is a group of particle painters that allow for the timed animation transitions for the grouped particle painters.

The direction that particles are emitted is controlled by the `Direction` items which consist of these components: `AngleDirection`, `PointDirection`, and `TargetDirection`.

There are only a few types of particle painters you can use, but they cover just about everything you would want to use them for. Particle types available in Qt Quick are as follows:

- `CustomParticle`: A particle based on OpenGL shader
- `ImageParticle`: A particle based on an image file
- `ItemParticle`: A particle based on a QML Item

`ImageParticle` is probably the most common and easiest to use and can be made from any image that QML has support for. If there are going to be numerous particles, it might be best to use small and optimized images.

Let's examine a simple `ItemParticle` animation. We will start by defining a `ParticleSystem` component with a child `ItemParticle` animation that is defined as a transparent `Rectangle` element with a small green border and a radius of 65, which means it appears as a green circle.

There are actually two type of emitters—the standard `Emitter` type, but also a special `TrailEmitter` type, which is derived from the `Emitter` item, but emits its particles from other particles instead of its bounding area.

An `Emitter` item is defined with the `SystemParticle` component bound to its `system` property. For the `velocity` property of the `Emitter` item, we use `AngleDirection`. `AngleDirection` directs the particles emitted at a certain angle.

Angles in QML elements work in a clockwise fashion, starting at the right-hand side of an element. Here's the representation of it:

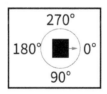

For example, setting an `AngleDirection` of 90 would make the particles move downward.

Let's dig into a particle example:

 The source code can be found on the Git repository under the `Chapter03-1` directory, in the `cp3` branch.

1. We start by defining a `ParticleSystem`:

```
ParticleSystem {
    id: particelSystem
    anchors.fill: parent
```

2. We add an `ItemParticle` and define the `delegate` to be a transparent `Rectangle`. We define a `radius`, which gives it rounded corners, and designate it to have a small green border:

```
ItemParticle {
    delegate: Rectangle {
        height: 30; width: 30
        id: particleSquare
        color: "transparent"
        radius: 65
        border.color: "green"
        border.width: 4
    }
}
}
```

3. We define an `Emitter` and assign it to the `ParticleSystem`:

```
Emitter {
    id: particles
    system: particleSystem
    anchors { horizontalCenter: parent.horizontalCenter; }
    y: parent.height / 2
    width: 10
    height: 10
    lifeSpan: 5000
    velocityFromMovement: 60
    sizeVariation: 15
    emitRate: 50
    enabled: false
```

4. We give the `Emitter` an `AngleDirection velocity` to add some variation in the direction:

```
velocity: AngleDirection {
    angle: 90
    magnitude: 150
    angleVariation: 25
    magnitudeVariation: 50
}
}
```

At this point, the app looks like this:

Let's see how the emitter would look when it is not centered:

1. We bind the `Emitter` property, called `enabled`, to the value of `false` in order to stop the particles being constantly emitted.
2. We then bind the `burst` property to animate a pulse of 25 particles with a mouse click like this:

```
MouseArea {
    id: mousey
    anchors.fill: parent
```

```
onClicked: {particles.burst(25) }
hoverEnabled: true
}
```

The properties of the `Emitter` component are attributes of the particles at the start of the animation.

3. We bind the `Emitter` property's x and y properties to the mouse position:

```
y: mousey.mouseY
x: mousey.mouseX
```

4. We can also remove the `horizontalCenter` anchor as well, unless you want the particle burst start to always be centered horizontally.

This image shows the `Emitter` when it's not centered horizontally:

To influence the particles as they get beamed out into the scene, you need an `Affector`. Let's take a look at how to use an `Affector` in the next section.

Affectors

An affector is an attribute that affects the way particles are streamed. There are a few types of `affectors` to choose from:

- `Age`: Will terminate particles early
- `Attractor`: Attracts particles toward a point
- `Friction`: Slows a particle proportional to its velocity
- `Gravity`: Applies acceleration at an angle
- `Turbulence`: Applies noise in a fluid manner
- `Wander`: Random particle trajectory

There are also `GroupGoal` and `SpriteGoal` `affectors`.

`Affectors` are optional but add their bling to the particles after they get emitted.

Let's examine one way to use these items.

1. We add a `Turbulence` item as a child to the `ParticleSystem` component. The particles will now fly around randomly, like falling leaves being blown around in the wind:

```
Turbulence {
    anchors.fill: parent
    strength: 32
}
```

2. You can have more than one affector. Let's add some `Gravity`, as well! We will make this `Gravity` go upward. `Gravity` is kind of like giving an item some weight in a certain direction:

```
Gravity {
    anchors.fill: parent
    angle: 270
    magnitude: 4
}
```

Here is what our example of `Turbulence` circles looks like:

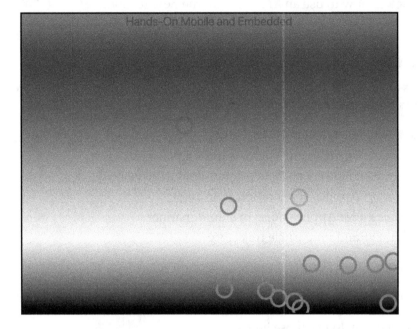

You can try the Qt for WebAssembly version here at `https://lpotter.github.io/particles/ch3-1.html`.

We can also cause the particles to flow in a particular direction, or act in a particular shape.

Shapes and directions

Shapes are a way that can be used to affect how Affectors act upon a certain area.

- `EllipseShape`: Acts on ellipse shaped area
- `LineShape`: Acts on a line
- `MaskShape`: Acts on an image shaped area
- `RectangleShape`: Acts on a rectangle area

Particles can have a velocity in a certain direction. There are three ways to direct particles:

- `AngleDirection`
- `PointDirection`
- `TargetDirection`

From the point of emissions, `AngleDirection` has four properties—`angle`, `angleVariation`, `magnitude`, and `magnitudeVariation`. As I mentioned previously, angles are measured in degrees clockwise, starting to the right of the `Emitter` item. The `magnitude` property specifies the velocity of movement in pixels per second.

`PointDirection` will direct a `velocity` property to a certain point in the scene, or off the screen, if you like. It takes the `x`, `y`, `xVariation`, and `yVariation` properties.

With `TargetDirection`, you can instruct particles to be emitted toward a target item, or a targeted `x`, `y` point. `TargetDirection` has a new property called `proportionalMagnitude`, which makes the `magnitude` and `magnitudeVariation` properties operate as a multiple of the distance between starting point and target point per second.

Particles can be quite fun and add a sci-fi element to an application. It takes some experimentation to get them to perform as you see in your mind, as there is a great randomness to them.

Now, let's look at adding some other types of effects for graphics.

Graphical effects for Qt Quick

When you usually think of effects such as blur, contrast, and glow, you might think of image editing software, as they tend to be applied those effects to images. Qt Graphical Effects can apply those same types of effects to QML UI components.

If you use the Qt Quick Scene Graph software renderer, these will not be available or usable, as this does not support the effects.

Qt Graphical Effects come in a variety of types, each with various sub-effects:

- `Blend`
- `Color:`
    - `BrightnessContrast`
    - `ColorOverlay`

- Colorize
- Desaturate
- GammaAdjust
- HueSaturation
- LevelAdjust
- Gradients:
  - ConicalGradient
  - LinearGradient
  - RadialGradient
- Displace
- DropShadows:
  - DropShadow
  - InnerShadow
- Blurs:
  - FastBlur
  - GaussianBlur
  - MaskedBlur
  - RecursiveBlur
- MotionBlurs:
  - DirectionalBlur
  - RadialBlur
  - ZoomBlur
- Glows:
  - Glow
  - RectangularGlow
- Masks:
  - OpacityMask
  - ThresholdMask

Now, let's move on to how DropShadow, being one of the most useful effects, works.

DropShadow

A `DropShadow` effect is something you can use to make things stand out and look more alive. It's usefulness is that it will give depth to otherwise flat objects.

We can add a `DropShadow` effect to a `Text` item from our last example. The `horizontalOffset` and `verticalOffset` properties characterize where the shadow will be perceived as being positioned upon the scene. The `radius` property describes the focus of the shadow, while the `samples` property determines the number of samples per pixel when blurring.

Use the following code to add a `DropShadow` and apply it to the `Text` component:

```
Text {
    id: textLabel
    text: "Hands-On Mobile and Embedded"
    color: "purple"
    font.pointSize: 20
    anchors.top: parent.top
    anchors.horizontalCenter: parent.horizontalCenter
}
DropShadow {
    anchors.fill: textLabel
    horizontalOffset: 2
    verticalOffset: 2
    radius: 10
    samples: 25
    color: "white"
    source: textLabel
}
```

The source code can be found on the Git repository under the `Chapter03-2` directory, in the `cp3` branch.

Here, you can see the letters now have a white shadow under them:

Hands-On Mobile and Embedded

It also has a `spread` property that controls the sharpness of the shadow. That is still a bit difficult to read, so let's try something else. How about a `Glow` effect?

Glow

Glow is an effect that produces a diffused color around the object by using the following code:

```
Glow {
    anchors.fill: textLabel
    radius: 10
    samples: 25
    color: "lightblue"
    source: textLabel
}
```

The effect is shown in the following screenshot. Notice the nice light blue glow:

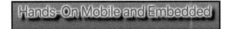

That's more like it! We can even give the Glow effect its own shadow! Change the DropShadow, anchors.fill, and source properties to glow:

```
DropShadow {
    anchors.fill: glow
    horizontalOffset: 5
    verticalOffset: 5
    radius: 10
    samples: 25
    color: "black"
    source: glow
}
```

Let's make the horizontalOffset and verticalOffset properties a tad bigger too.

Our banner now looks like this:

DropShadows are great for making something stand out from the scene. Gradients are another effect to use.

Gradient

Gradients can grab a user's attention, draw them into a UI, and connect to their emotions.

Qt Graphical Effects have built-in support for three types of gradients—Conical, Linear, and Radial.

RadialGradient, or any QML gradient for that matter, is made up of a series of GradientStop items, which specify the color and where to start it in the gradient cycle, the number zero being at the beginning, and one being at the end point.

Here's the code for representing a RadialGradient:

```
Item {
    width: 250; height: 250
    anchors.horizontalCenter: parent.horizontalCenter
    anchors.verticalCenter: parent.verticalCenter
    RadialGradient {
        anchors.fill: parent
        gradient: Gradient {
            GradientStop { position: 0.0; color: "red" }
            GradientStop { position: 0.3; color: "green" }
            GradientStop { position: 0.6; color: "purple" }
        }
    }
}
```

The source code can be found on the Git repository under the Chapter03-4 directory, in the cp3 branch.

The following is the pictorial presentation of our RadialGradient:

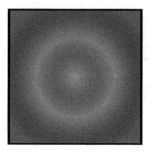

This `RadialGradient` uses three `GradientStop` items to tell the gradient where a certain color should start from. The `position` property is a `qreal` value from `0.0` to `1.0`; although having a number larger than `1.0` will not give an error, it will simply not be drawn in the bounding item.

Using the same color-stop schemes as the `RadialGradient`, we see how the `LinearGradient` and `ConicalGradient` look.

The following is a representation of `LinearGradient`:

The following is a representation of `ConicalGradient`:

You can see the differences between each of these gradients.

Blur

Blur effects can help de-emphasize or bring action to a static image. The fastest Blur effect would be the aptly named FastBlur effect, but the GaussianBlur effect is the highest quality, and, consequently, the slowest to render.

All the Blur effects have radius, samples, and source properties. Radius represents the distance of pixels that will affect the Blur effect, with a higher number increasing the Blur effect. The samples property represents the number of samples per pixel used when the effect is applied. A higher number means better quality, but a slower render time. Source is the source item that the Blur effect will be applied to.

Displace is a type of a Blur effect, but with more of a possible watermark-type effect. The displacementSource property is the item that is being interposed onto the source item. The displacement property is a qreal value between -1.0 and 1.0, with 0 meaning there is no displacement of pixels.

Summary

Qt Quick offers graphical and special effects that are very easy to start using. Particles, in particular, are great for gaming apps. You now know how to use the ParticleSystem to emit an ImageParticle at a particular angle using AngleDirection. We examined how Affectors such as Turbulence will affect an Emitter by adding variation to the particle stream.

Gradients, Glow, and DropShadows are useful for bringing emphasis to certain items. The Blur effects are used to simulate movement action or to add your watermark to images.

In the next chapter, we delve into using something now ubiquitous on mobile phones—touch input. I will also touch upon (pun intended) using other forms of inputs, such as what to do when there is no hardware keyboard and your app is what gets booted into.

4
Input and Touch

Not all devices have a readily available keyboard. With a touchscreen device, users can easily use buttons and other **User Interface** (**UI**) features. What do you do when there is no keyboard or mouse, like on a kiosk or interactive signage? Virtual keyboards and touch interaction define mobile and embedded applications these days.

In this chapter, we will cover the following topics:

- We will discover Qt's graphical solutions to incorporating user input.
- The reference Qt Virtual Keyboard will be examined.
- We will demonstrate Touch components, such as `TouchPoints`, `Flickable`, and `PinchArea`.

What to do when there's no keyboard

Dude, where's my keyboard?

Computer information kiosks and cars do not usually come with keyboard inputs. They use virtual inputs, such as a virtual keyboard, voice inputs, or even gesture recognition.

People at the Qt Company have created a virtual input method they named **Qt Virtual Keyboard** (**QtVK**). It's more than just an onscreen keyboard, as it also has handwriting recognition. It is available under a commercial license as well as the open source GPL version 3.

There are other virtual keyboards that will work with Qt apps. On a desktop computer that also has a touchscreen, such as a two-in-one laptop, the system might already have a virtual keyboard. These should work as an input method for Qt apps, although they may or may not automatically pop up when the user wants to input into a text area.

There are two ways to integrate Qt's Virtual Keyboard:

Desktop system	Fully integrated into applications
Application	Qt Widget apps: Set environmental `QT_IM_MODULE=qtvirtualkeyboard` variable Qt Quick: Use `InputPanel` in your application

I have a Raspberry Pi setup here for Boot to Qt, which is fully integrated into the Qt Creator, so I can build and run Qt apps on the Raspberry Pi from the Qt Creator. You can also grab the sources and build it yourself from `git://code.qt.io/qt/qtvirtualkeyboard.git`.

To build the QtVK, download the following source:

```
git clone git://code.qt.io/qt/qtvirtualkeyboard.git
```

QtVK build can be configured by `qmake`, using `CONFIG+=<configuration>` and the following configuration options:

- `lang <code>`
    - `form of language_country`
        - Language is lowercase, a two-letter language code
        - Country is uppercase, a two-letter country code

- `lang-all`
- `handwriting`
    - Handles custom engines
- Arrow-key navigation

 For example, to configure only Australian-English and add handwriting support, you would run `qmake CONFIG+=lang-en_AU CONFIG+=handwriting` and then `make && make install`.

There are many other configurations available. You can disable layouts for creating a custom layout and desktop integration, among other configurations.

QtVK can be used in C++ or QML. Let's start by using Qt Quick by ticking the **Use Qt Virtual Keyboard** in the Qt Creator project wizard when creating a new **Qt Quick Application** from a template:

This is the boiler plate code you get when you use Boot to Qt for Device Creation:

```
InputPanel {
    id: inputPanel
    z: 99
    x: 0
    y: window.height
    width: window.width
    states: State {
        name: "visible"
        when: inputPanel.active
        PropertyChanges {
            target: inputPanel
            y: window.height - inputPanel.height
        }
    }
    transitions: Transition {
        from: ""
        to: "visible"
        reversible: true
        ParallelAnimation {
            NumberAnimation {
                properties: "y"
                duration: 250
                easing.type: Easing.InOutQuad
            }
        }
    }
}
```

The source code can be found on the Git repository under the Chapter04-1 directory, in the cp4 branch.

Let's add something that takes text as input, such as a TextField element:

```
TextField {
    anchors {
        bottom: inputPanel.top
        top: parent.top
        right: parent.right
        left: parent.left
    }
    placeholderText: "Enter something"
}
```

The anchors here are used to resize this TextField element when QtVK is opened automatically when the user taps on TextField.

Here is what this should look like:

Implementing a touch screen can have many benefits, which we can do by using the Qt event loop. Let's look in detail at using touch screens as an input.

Using touch input

Touch screens are ubiquitous these days. They are everywhere. While essential in a mobile phone or tablet, you can also get a laptop or desktop computer with one. Refrigerators and cars also commonly have touchscreens. Knowing how to utilize these in your Qt app is also essential.

On mobile phone and tablet platforms, touchscreen support comes from the system and is often built-in. If you are creating your own embedded device, you will most likely need to tell Qt how to use the touchscreen. Qt has support for various touchscreen systems on embedded devices.

QEvent

QEvent is the way to get access to the touch input events in C++. It comes through an event filter you can add to your application. There are a few different ways to access this data.

We can use an event filter or an event loop. We will start by looking at an event filter.

Event filter

One way you can access the event loop is by using an event filter. You first need to call the following function:

```
qApp->installEventFilter(this);
```

> The source code can be found on the Git repository under the Chapter04-2 directory in the cp4 branch.

You then need to override the function named eventFilter(QObject* obj, QEvent* event), which returns a bool value:

```
bool MainWindow::eventFilter(QObject* obj, QEvent* event);
```

You will then receive any and all events. You can also handle these touch events by using the following:

- QEvent::TouchBegin
- QEvent::TouchCancel
- QEvent::TouchEnd
- QEvent::TouchUpdate

Using a switch statement in a eventFilter is an effective way to go through different options:

```
bool MainWindow::eventFilter(QObject* obj, QEvent* event)
{
    switch(event->type()) {
        case QEvent::TouchBegin:
        case QEvent::TouchCancel:
        case QEvent::TouchEnd:
        case QEvent::TouchUpdate:
            qWarning("Touch event %d", event->type());
            break;
        default:
            break;
    };
    return false;
}
```

Be sure to pass these events on to the parent class unless you need to intercept them. To not pass these on, return true. Using an event loop is another way to access events. Let's take a look.

Event loop

To use the event loop, you need to override event(QEvent *ev):

```
bool MainWindow::event(QEvent *ev)
{
  switch (ev->type()) {
    case QEvent::TouchBegin:
      qWarning("TouchBegin event %d", ev->type());
        break;
    case QEvent::TouchEnd:
      qWarning("TouchEnd event %d", ev->type());
        break;
    case QEvent::TouchUpdate:
      qWarning("TouchUpdate event %d", ev->type());
```

```
        break;
    };}
```

You also need to add `setAttribute(Qt::WA_AcceptTouchEvents, true);` to the class constructor, otherwise your application will not receive touch events.

Let's take a look at how touchscreen support is handled in Qt and how you can use Qt to access a lower level of the touchscreen input stack.

Touchscreen support

Touchscreen support for Qt is done through the **Qt Platform Abstraction (QPA)** platform plugins.

Qt configure will auto-detect the correct platform and determine whether or not the development files are installed. If it finds the development files, it will use them.

Let's see how touchscreens work for various operating systems, starting out with mobile phones.

Windows, iOS and Android

On Windows, iOS and Android, touchscreens are supported through the Qt Event system.

Using the Qt event system and allowing the platform plugins to do the scanning and reading, we can use `QEvent` if we need access to those events.

Let's look at how we can access a low level of the input system using Qt on embedded Linux.

Linux

On the Linux operating system there are a variety of input systems that can be used with Qt.

Qt has built-in support for these types of touchscreen interfaces:

- `evdev`: an event device interface
- `libinput`: a library to handle input devices
- `tslib`: a typescript runtime library

We will start by learning about the Linux `evdev` system to read the device files directly.

evdev

Qt has built-in support for the `evdev` standard event-handling system for Linux and embedded Linux . This is what you will get by default if no other system is configured or detected. It handles keyboard, mouse, and touch. You can then use Qt as normal with respect to keyboard, touch, and mouse events.

You can assign startup parameters, such as device file path and default rotation of the screen, like this:

```
QT_QPA_EVDEV_TOUCHSCREEN_PARAMETERS=/dev/input/input2:rotate=90
```

Other parameters available are `invertx` and `inverty`. Of course, you do not need to reply on Qt for these input events, and can access them directly in the stack below Qt. I call them raw events, but they are really just reading the special Linux kernel device files.

Let's take a look at handling these `evdev` input events yourself while using Qt. This is low-level system file access, so you might need root or administrator permissions to run the applications that use it this way.

Input events on Linux are accessed through the kernel's `dev` nodes, typically found at `/dev/input`, but they could be anywhere under the `/dev` directory tree, depending on the driver. `QFile` should not be used for actually reading from these special device node files.

 QFile is not suited for reading Unix device node files. This is because QFile has no signals and the device node files report a size of zero and only have data when you read them.

The main `include` file to read input nodes is as follows:
```
#include <linux/input.h>
```

 The source code can be found on the Git repository under the `Chapter04-3` directory in the `cp4` branch.

You will want to scan the device files to detect which file the touchscreen produces. In Linux, these device nodes are named dynamically, so you need to use some other method to discern the correct file other than just the filename. So, you have to open the file and ask it to tell you its name.

We can use QDir and its filters to at least filter out some of the files we know are not what we are looking for:

```
QDir inputDir = QDir("/dev/input");
QStringList filters;
filters << "event*";
QStringList eventFiles = inputDir.entryList(filters,
QDir::System);
int fd = -1;
char name[256];
for (QString file : eventFiles) {
    file.prepend(inputDir.absolutePath());
    fd = ::open(file.toLocal8Bit().constData(), O_RDONLY|O_NONBLOCK);
if (fd >= 0) {
ioctl(fd, EVIOCGNAME(sizeof(name)), name);
::close(fd);
}
}
```

Be sure to include the O_NONBLOCK argument for open.

At this point, we have a list of the names for the different input devices. You might have to just guess which name to use and then do a String compare to find the correct device. Sometimes, the driver will have correct id information, which can be obtained using EVIOCGID like this:

```
unsigned short id[4];
ioctl(fd, EVIOCGID, &id);
```

Sometimes, you can detect certain features using EVIOCGBIT. This will tell us which buttons or keys the hardware driver supports. The touchscreen driver outputs a keycode of 0x14a (BTN_TOUCH) when you touch it, so we can use this to detect which input event will be our touchscreen:

```
bool MainWindow::isTouchDevice(int fd)
{
    unsigned short id[4];
    long bitsKey[LONG_FIELD_SIZE(KEY_CNT)];
    memset(bitsKey, 0, sizeof(bitsKey));
    ioctl(fd, EVIOCGBIT(EV_KEY, sizeof(bitsKey)), bitsKey);
    if (testBit(BTN_TOUCH, bitsKey)) {
```

```
            return true;
        }
        return false;
    }
```

We can now be fairly certain that we have the proper device file. Now, we can set up a QSocketNotifier object to notify us when that file is activated, and then we can read it to get the X and Y values of the touch. We use the QSocketNotifier class because we cannot use QFile, as it doesn't have any signals to tell us when the Linux device files get changed, so this makes it much easier:

```
int MainWindow::doScan(int fd)
{
    QSocketNotifier *notifier
            = new QSocketNotifier(fd, QSocketNotifier::Read,
this);
    auto c = connect(notifier,  &QSocketNotifier::activated,
                        [=]( int /*socket*/ ) {
        struct input_event ev;
        unsigned int size;
        size = read(fd, &ev, sizeof(struct input_event));
        if (size < sizeof(struct input_event)) {
            qWarning("expected %u bytes, got %u\n", sizeof(struct
input_event), size);
            perror("\nerror reading");
            return EXIT_FAILURE;
        }
        if (ev.type == EV_KEY && ev.code == BTN_TOUCH)
            qWarning("Touchscreen value: %i\n", ev.value);
        if (ev.type == EV_ABS && ev.code == ABS_MT_POSITION_X)
            qWarning("X value: %i\n", ev.value);
         if (ev.type == EV_ABS && ev.code == ABS_MT_POSITION_Y)
            qWarning("Y value: %i\n", ev.value);
          return 0;
    });
  return true;
}
```

We also use the standard read() function instead of QFile to read this.

The BTN_TOUCH event value tells us when the touchscreen was pressed or released.

The ABS_MT_POSITION_X value will be the touchscreen's X position, and the ABS_MT_POSITION_Y value will be the Y position.

There is a library that can be used to do the very same thing, which might be a little easier.

libevdev

When you use the library `libevdev`, you won't have to access such low level filesystem functions like a `QSocketNotifier` and read files yourself.

To use `libevdev`, we start by adding to the `LIBS` entry in our projects `.pro` file.

```
LIBS += -levdev
```

 The source code can be found on the Git repository under the `Chapter04-4` directory in the `cp4` branch.

This allows `qmake` to set up proper linker arguments. The `include` header would be as follows:

```
#include <libevdev-1.0/libevdev/libevdev.h>
```

We can borrow the initial code to scan the directory for device files from the preceding code, but the `isTouchDevice` function gets cleaner code:

```
bool MainWindow::isTouchDevice(int fd)
{
    int rc = 1;
    rc = libevdev_new_from_fd(fd, &dev);
    if (rc < 0) {
        qWarning("Failed to init libevdev (%s)\n", strerror(-rc));
        return false;
    }
    if (libevdev_has_event_code(dev, EV_KEY, BTN_TOUCH)) {
        qWarning("Device: %s\n", libevdev_get_name(dev));
        return true;
    }
    libevdev_free(dev);
    return false;
}
```

Libevdev has the nice `libevdev_has_event_code` function that can be used to easily detect whether or not the device has a certain event code. This is just what we needed to identify the touchscreen! Notice the `libevdev_free` function, which will free the memory being used that we do not need.

The doScan function loses the call to read, but substitutes a call to libevdev_next_event instead. It can also output a nice message with the actual name of the event code by calling libevdev_event_code_get_name:

```cpp
int MainWindow::doScan(int fd)
{
    QSocketNotifier *notifier
            = new QSocketNotifier(fd, QSocketNotifier::Read,
this);
    auto c = connect(notifier,  &QSocketNotifier::activated,
                    [=]( int /*socket*/ ) {
        int rc = -1;
        do {              struct input_event ev;
            rc = libevdev_next_event(dev,
LIBEVDEV_READ_FLAG_NORMAL, &ev);
            if (rc == LIBEVDEV_READ_STATUS_SYNC) {
                while (rc == LIBEVDEV_READ_STATUS_SYNC) {
                    rc = libevdev_next_event(dev, LIBEVDEV_READ_FLAG_SYNC,
&ev);
                }
            } else if (rc == LIBEVDEV_READ_STATUS_SUCCESS) {
                if ((ev.type == EV_KEY && ev.code == BTN_TOUCH) ||
                        (ev.type == EV_ABS && ev.code ==
ABS_MT_POSITION_X) ||
                        (ev.type == EV_ABS && ev.code ==
ABS_MT_POSITION_Y)) {
                    qWarning("%s value: %i\n",
libevdev_event_code_get_name(ev.type, ev.code), ev.value);
                }
            }
        } while (rc == 1 || rc == 0 || rc == -EAGAIN);
        return 0;
    });
    return 0;
}
```

The library libinput also uses evdev, and is a bit more up to date than the others.

libinput

The `libinput` library is the input handling for Wayland compositors and X.Org window system. Wayland is a display server protocol a bit like a newer version of the ancient Unix standard X11. `Libinput` depends on `libudev` and supports the following input types:

- `Keyboard`: Standard hardware keyboard
- `Gesture`: Touch gestures
- `Pointer`: Mouse events
- `Touch`: Touchscreen events
- `Switch`: Laptop lid switch events
- `Tablet`: Tablet tool events
- `Tablet pad`: Tablet pad events

The `libinput` library has a build-time dependency upon `libudev` ; therefore, to configure Qt, you will need `libudev` as well as the `libinput` development files or packages installed. If you need hardware keyboard support, the `xcbcommon` package is also needed.

Yet another touch library is `tslib`, which is specifically used in embedded devices, as it has a small filesystem footprint and minimal dependencies.

Tslib

`Tslib` is a library used to access and filter touchscreen events on Linux devices; it supports multi-touch and Qt has support for using it. You will need to have the `tslib` development files installed. Qt will auto-detect this, or you can explicitly configure Qt with the following:

```
configure -qt-mouse-tslib
```

It can then be enabled by setting `QT_QPA_EGLFS_TSLIB` or `QT_QPA FB_TSLIB` to 1. You can change the actual device file path by setting the environmental variable called `TSLIB_TSDEVICE` to the path of the device node like this:

```
export TSLIB_TSDEVICE=/dev/input/event4
```

Let's now move on to see how we can use higher level APIs in Qt to utilize the touchscreen.

Using the touchscreen

There are two ways the Qt backend uses the touchscreen. The events come in as a mouse using a point with `click` and `drag` events, or as multi-point touch-to-handle gestures, such as `pinch` and `swipe`. Let's get a better understanding about multi-point touch.

MultiPointTouchArea

As I have mentioned earlier, to use multi-point touchscreens in QML, there is the `MultiPointTouchArea` type. If you want to use gestures in QML, you either have to use `MultiPointTouchArea` and do it yourself, or use `QGesture` in your C++ and handle custom signals in your QML components.

 The source code can be found on the Git repository under the `/Chapter04-5` directory in the cp4 branch.

```
MultiPointTouchArea {
    anchors.fill: parent
    touchPoints: [
        TouchPoint { id: finger1 },
        TouchPoint { id: finger2 },
        TouchPoint { id: finger3 },
        TouchPoint { id: finger4 },
        TouchPoint { id: finger5 }
    ]
}
```

You declare the `touchPoints` property of `MultiPointTouchArea` with a `TouchPoint` element for each finger you want to deal with. Here, we are using five-finger points.

You can use the x and y properties to move things around:

```
Rectangle {
    width: 30; height: 30
    color: "green"
    radius: 50
    x: finger1.x
    y: finger1.y
}
```

You can also use touchscreen gestures in your app.

Qt Gestures

Gestures are a great way of utilizing user input. As I mentioned gestures in `Chapter 2`, *Fluid UI with Qt Quick*, I will mention the C++ API here, which is much more feature-rich than gestures in QML. Keep in mind that these are touchscreen gestures and not device or sensor gestures, which I will examine in a later chapter. `QGesture` supports the following built-in gestures:

- `QPanGesture`
- `QPinchGesture`
- `QSwipeGesture`
- `QTapGesture`
- `QTapAndHoldGesture`

QGesture is an event-based API, so it will come through the event filter, which means you need to re-implement your `event(QEvent *event)` widgets as the gesture will target your widget. It also supports custom gestures by subclassing `QGestureRecognizer` and re-implementing `recognize`.

To use gestures in your app, you need to first tell Qt that you want to receive touch events. If you are using built-in gestures, this is done internally by Qt, but if you have custom gestures, you need to do this:

```
setAttribute(Qt::WA_AcceptTouchEvents);
```

To accept touch events, you then need to call `QGraphicsItem::setAcceptTouchEvent(bool)` with `true` as the argument.

If you want to use unhandled mouse events for touch events, you can also set the `Qt::WA_SynthesizeTouchEventsForUnhandledMouseEvents` attribute.

You then need to define to Qt that you want to use certain gestures by calling the `grabGesture` function of your `QWidget` or `QGraphicsObject` class:

```
grabGesture(Qt::SwipeGesture);
```

> `QGesture` events are delivered to a specific `QWidget` class and not the current `QWidget` class that holds the focus like a mouse event would.

In your `QWidget` derived class, you need to re-implement the `event` function and then handle the `gesture` event when it happens:

```
bool MyWidget::event(QEvent *event)
{
    if (event->type() == QEvent::Gesture)
        handleSwipe();
    return QWidget::event(event);
}
```

Since we are handling only one `QGesture` type, we know it is our target gesture. You can check whether or not this event is caused by a certain gesture by checking for its `pointer` using the `gesture` function that is defined as follows:

```
QGesture * QGestureEvent::gesture(Qt::GestureType type) const
```

This can be implemented by the following:

```
if (QGesture *swipe = event->gesture(Qt::SwipeGesture))
```

If the `QGesture` object called `swipe` is `nullptr`, then this event is not our target gesture.

It is also a good idea to check on the gesture's `state()`, which can be one of the following:

- `Qt::NoGesture`
- `Qt::GestureStarted`
- `Qt::GestureUpdated`
- `Qt::GestureFinished`
- `Qt::GestureCanceled`

You can create your own gesture using `QGestureRecognizer` by sub-classing `QGestureRecognizer` and re-implementing `recognize()`. This is where most of the work will be, as you will need to detect your gesture and are more likely detect what is not your gesture. Your `recognize()` function will need to return one of the values of the enum value, `QGestureRecognizer::Result`, which can be any of the following:

- `QGestureRecognizer::Ignore`
- `QGestureRecognizer::MayBeGesture`
- `QGestureRecognizer::TriggerGesture`
- `QGestureRecognizer::FinishGesture`
- `QGestureRecognizer::CancelGesture`
- `QGestureRecognizer::ConsumeEventHint`

There are heaps of edge cases you need to handle here to discern exactly what is and what is not your gesture. Do not be afraid if this function is complicated or long.

Another form of input that is becoming more popular is using your voice. Let's look at that next.

Voice as input

Voice recognition and Qt has been around for a while now. IBM's ViaVoice was ported to KDE and was being ported to Trolltech's phone software suite Qtopia at the time that I became the Qtopia Community Liaison in 2003. Consequently, it was worked on by the same developer who later dreamed-up what became Qt Quick. While the concept has essentially stayed the same, the technology has gotten better, and now you can find voice control in many different devices, including automobiles.

There are many competing systems, such as Alexa, Google Voice, Cortana, and Siri, as well as some open source APIs. Combined with voice search, voice input is an invaluable tool.

At the time of writing, Qt Company and one of its partners, **Integrated Computer Solutions (ICS)**, have announced that they are working together on integrating the Amazon Alexa system with Qt. I have been told that it will be called `QtAlexaAuto` and be released under the lgpl v3 license. Since this has not been released at the time of writing, I cannot go into very much detail about how to use this implementation just yet. I am sure that, if or when it gets released, the API will be quite easy to use.

Amazon's **Alexa Voice Service (AVS) Software Development Kit (SDK)** works on Windows, Linux, Android and MacOS. You can even use a mic array component, such as the MATRIX creator with the Raspberry Pi. Siri works on iOS and MacOS. Cortana works on Windows.

While none of these voice systems are integrated into Qt, they can be used with a custom integration. It is worth looking into them, depending on what your application will be doing and what device it will run on.

Alexa, Google Assistant, and Cortana have C++ APIs, and Siri can be used as well with its Objective-C API:

- **Alexa:** `https://github.com/alexa/avs-device-sdk.git`
- **Google Assistant:** `https://developers.google.com/assistant/sdk/guides/library/python/embed/install-sample`

- **Cortana:** `https://developer.microsoft.com/en-us/cortana`
- **Siri:** `https://developer.apple.com/sirikit/`

QtAlexaAuto

QtAlexaAuto is a module created by Integrated Computer Solutions (ICS) and the Qt Company to enable the use of Amazon's Alexa Voice Service (AVS) from within Qt and QtQuick applications. You can use Raspberry Pi, Linux, Android, or other machines to prototype an application that uses voice as input.

At the time of writing this book, QtAlexaAuto is yet to be released, so you will need to search the internet for the url to download the source code. Some things may have changed in the official release from what it written in this book.

You will need to download, build and install the following SDKs from Amazon:

- AVS Device SDK: git clone -b v1.9 `https://github.com/alexa/avs-device-sdk`
- Alexa Auto SDK: git clone -b 1.2.0 `https://github.com/alexa/aac-sdk`

To build these you should follow instructions for your platform from this URL:

`https://github.com/alexa/avs-device-sdk/wiki`

The basic steps for building QtAlexaAuto are the following:

1. Sign-up for an Amazon developer account, and register a product. `https://developer.amazon.com/docs/alexa-voice-service/register-a-product.html`
2. Add your clientID and productID
3. Install the requirements mentioned in the Alexa wiki
4. Apply patches as detailed in install_aac_sdk.md
5. Build and install AVS Device SDK
6. Build and install Alexa Auto SDK
7. Edit AlexaSamplerConfig.json
8. Build QtAlexaAuto

There are a few patches you need to apply. Luckily there are instructions in install_aac_sdk.md, which instruct you on how to apply the patches from aac-sdk to the avs-device-sdk.

The file AlexaSamplerConfig.json needs to be edited and renamed to AlexaClientSDKConfig_new_version_linux.json

The file then needs to be put into the directory where you are running the example from.

The QtAlexaAuto main QML component is named alexaApp, which corresponds to the QtAlexaAuto class.

When you run the example app, you will need to sign in to your Amazon developer account and link this application by entering a code the application gives you when you start it for the first time. These can be provided to the user by calling acctLinkUrl() and acctLinkCode(), or in QML by the alexaApp properties named accountLinkCode and accountLinkUrl.

Once this is linked to an account, you use voice input and the Alexa Voice Service by tapping on the button.

The function that runs when the user presses the talk button is tapToTalk(), and the startTapToTalk signal gets emitted.

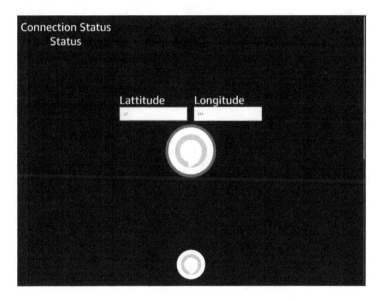

AVS has a notion of a RenderTemplate, which gets passed from the service so the application will be able to show the user visual information about the response. QtAlexaAuto handles Weather, media player templates, as well as some generic multipurpose templates. The RenderTemplate gets emitted as JSON documents and shown in the example application using QML components, which then parse and display the data.

This was just a quick look at QtAlexaAuto, as I did not have enough time to really dig into this new API before the publication of this book.

Summary

User input is important, and there are various ways for users to interact with applications. If you are creating your own embedded device, you will need to decide what input methods to use. Touchscreens can increase usability because touching things is a very natural thing to do. Babies and even cats can use touchscreen devices! Gestures are a fantastic way to use touch input and you can even develop custom gestures for your application. Voice input is taking off right now. Whilst adding support for it might take a little work, it can be the right thing to do on some devices that require hands-free usage.

In the following chapter, we will learn about networking and its features.

2
Section 2: Networking, Connectivity, Sensors, and Automation

In this section you will learn about additional features and mobile APIs to extend applications beyond their device. Reading data from remote sensors, and using APIs to communicate with machines will be discussed. QNetworkReply, QNetworkRequest, QDnsLookup, QHostInfo, QLocalServer, and QTcpSocket APIs will be covered. We'll also discuss QNetworkSession and QNetworkConfiguration which is used to see available Wi-Fi networks nearby.

This section comprises of the following chapters:

- Chapter 5, *Qt Network for Communication*
- Chapter 6, *Connectivity with Qt Bluetooth LE*
- Chapter 7, *Machines Talking*
- Chapter 8, *Where Am I? Location and Positioning*

Qt Network for Communication

5

Networking is almost as important to mobile devices as the device being mobile. Without networking, data would have to be physically moved from one place to another. Luckily, Qt has extensive networking features in `QNetwork`. In this chapter, we will discuss the following APIs:

- `QNetworkReply`
- `QNetworkRequest`
- `QDnsLookup`
- `QHostInfo`
- `QLocalServer`
- `QTcpSocket`

To show available Wi-Fi networks that are nearby, we will also go over the following:

- `QNetworkSession`
- `QNetworkConfiguration`

You will also learn how to use Qt APIs for standard networking tasks, such as **Domain Name Service** (**DNS**) lookups, download and upload files, and how to use Qt's socket classes for communication.

High level – request, reply, and access

Networking in Qt is quite feature-rich. Networking in Qt Quick is more behind the scenes than in your face. In **Qt Modeling Language** (**QML**), you can download remote components and use them in your application, but any other arbitrary download or network functionality you will have to bake yourself in the C++ backend or use JavaScript.

Even though `QNetworkRequest`, `QNetworkReply`, and `QNetworkAccessManager` are all used to make network requests, let's split them up and see how to use them.

QNetworkRequest

`QNetworkRequest` is a part of the access functionality. It constructs a `request`, which can be one of the following verbs:

- GET: `get(...)`
- POST: `post(...)`
- PUT: `put(...)`
- DELETE: `deleteResource(...)`
- HEAD: `head(...)`

You can also send custom verbs using `sendCustomRequest`, which takes the custom verb as a `QByteArray` argument.

Headers can be set as known headers using `setHeader` and can be one of the following:

- `ContentDispositionHeader`
- `ContentTypeHeader`
- `ContentLengthHeader`
- `LocationHeader`
- `LastModifiedHeader`
- `CookieHeader`
- `SetCookieHeader`
- `UserAgentHeader`
- `ServerHeader`

Raw or custom headers can be set with `setRawHeader`. HTTP attributes can help to control the request cache, redirect, and cookies. They can be set with, you guessed it, `setAttribute`.

Let's put this into the following code.

 The source code can be found on the Git repository under the
`Chapter05-1` directory, in the `cp5` branch.

To use the networking module, in the `.pro` project, add `network` to the `QT` variable as
follows:

```
QT += network
```

We can now use Qt Networking.

`QNetworkRequest` is what needs to be used to request operations from the network such
as `get` and `put`.

A simple implementation looks like this:

```
QNetworkRequest request;
request.setUrl(QUrl("http://www.example.com"));
```

`QNetworkRequest` can also take `QUrl` as its argument. `QNetworkRequest` is not based on
`QObject`, so it has no parent, nor does it have any of its own signals. All communication is
done through `QNetworkAccessManager`.

The one signal you want to connect to is the `finished` signal.

Suppose I have some form data I need to transfer; I would need to add a standard header
with `setHeader`. I could also add the following custom header I call `X-UUID`:

```
request.setHeader(QNetworkRequest::ContentTypeHeader, "application/x-www-
form-urlencoded");
request.setRawHeader(QByteArray("X-UUID"),
QUuid::createUuid().toByteArray());
```

Now that we have a viable `QNetworkRequest`, we need to send it to
`QNetworkAccessManager`. Let's take a look at how we can do that.

QNetworkAccessManager

Bring in the manager—QNetworkAccessManager (**QNAM**). It is used to send and receive asynchronous requests over a network. Usually, there is one instance of QNAM in an application, as here:

```
QNetworkAccessManager *manager = new QNetworkAccessManager(this);
```

At its simplest, you can make a QNAM request using the get, put, post, deleteResource, or head functions.

QNAM uses signals to transfer data and request information and the finished() signal is used to signal when a request has finished.

Let's add a signal handler for that, as follows:

```
connect(manager, &QNetworkAccessManager::finished,
        this, &MainWindow::replyFinished);
```

This would call your replyFinished slot with the data and headers within the QNetworkReply argument, as follows:

```
void MainWindow::replyFinished(QNetworkReply *reply)
{
    if (reply->error())
        ui->textEdit->insertPlainText( reply->errorString());
    else {
        QList<QByteArray> headerList = reply->rawHeaderList();
        ui->textEdit->insertPlainText(headerList.join("\n") +"\n");
        QByteArray responsData = reply->readAll();
        ui->textEdit->insertHtml(responsData);
    }
}
```

Then, call the get method on QNetworkAccessManager as follows:

```
manager->get(request);
```

It's as simple as that to download something! QNAM will work its magic and download the URL.

It is also just as easy a method to create a file upload. Of course, your web server needs to support the put method, as follows:

```
QFileDialog dialog(this);
dialog.setFileMode(QFileDialog::AnyFile);
QString filename = QFileDialog::getOpenFileName(this, tr("Open File"),
```

```
QDir::homePath());

    if (!filename.isEmpty()) {
        QFile file(filename);
        if (file.open(QIODevice::ReadOnly | QIODevice::Text)) {
            QByteArray fileBytes = file.readAll();
            manager->put(request, fileBytes);
        }
    }
```

The source code can be found on the Git repository under the `Chapter05-2` directory, in the `cp5` branch.

If you need to send some query parameters in the URL, you can use `QUrlQuery` to construct the `form` query data, and then send the `request` as follows:

```
QNetworkRequest request;
QUrl url("http://www.example.com");

QUrlQuery formData;
formData.addQueryItem("login", "me");
formData.addQueryItem("password", "123");
formData.addQueryItem("submit", "Send");
url.setQuery(formData);
request.setUrl(url);
manager->get(request);
```

Form data can be uploaded with the `post` function as a `QByteArray` as follows:

```
QByteArray postData;
postData.append("?login=me&password=123&submit=Send");
manager->post(request, postData);
```

To send a multipart form data, such as form data and an image, you can use `QHttpMultiPart` as follows:

```
QFile *file = new QFile(filename);
    if (file->open(QIODevice::ReadOnly)) {
        QByteArray fileBytes = file->readAll();
         QHttpMultiPart *multiPart =
            new QHttpMultiPart(QHttpMultiPart::FormDataType);

        QHttpPart textPart;
        textPart.setHeader(QNetworkRequest::ContentDispositionHeader,
QVariant("form-data; name=\"filename\""));
        textPart.setBody(filename.toLocal8Bit());

        QHttpPart filePart;
```

```
        filePart.setHeader(QNetworkRequest::ContentDispositionHeader,
QVariant("form-data; name=\"file\""));

        filePart.setHeader(QNetworkRequest::ContentTypeHeader,
QVariant("application/zip"));

        filePart.setBodyDevice(file);

        file->setParent(multiPart);

        multiPart->append(textPart);
        multiPart->append(filePart);

        manager->put(request, multiPart);
    }
```

Of course, none of these examples keeps track of the reply. QNetworkReply is returned by the get, post, and put methods of QNetworkAccessManager, which can be used to track download or upload progress or if there are any errors.

QNetworkReply

All calls to QNAM's get, post, and so on, will return QNetworkReply.

 You will need to delete this pointer, otherwise it will leak memory, but do not delete it in the finished signal handler. You can use deleteLater().

QNetworkReply has an interesting signal we would most likely need to handle. Let's start with the two most important—error and readyRead.

So, let's handle that QNetworkReply properly. Since we do not have the valid object beforehand, we need to connect the signals after the network request action. It seems a bit backward to me, but this is the way it needs to be and it works. The code is as follows:

```
QNetworkReply *networkReply = manager->get(request);
connect(networkReply, SIGNAL(finished()), this, SLOT(requestFinished()));
connect(networkReply, SIGNAL(error(QNetworkReply::NetworkError)),
    this,SLOT(networkReplyError(QNetworkReply::NetworkError)));
  connect(networkReply, SIGNAL(readyRead()), this, SLOT(readyRead()));
```

I am using the legacy style of signal connections, but you could and should write connections like the following because it allows compile time checking for syntax and other errors:

```
connect(networkReply, &QNetworkReply::error, this,
&MyClass::networkReplyError);
  connect(networkReply,
QOverload<QNetworkReply::NetworkError>::of(&QNetworkReply::error),this,
&MyClass::networkReplyError);
connect(networkReply, &QNetworkReply::finished, this,
&MyClass::requestFinished);
connect(networkReply, &QNetworkReply::readyRead, this,
&MyClass::readyRead);
```

So, now we have done a request and are waiting for a reply from the server. Let's look at the signal handlers one by one.

`error(QNetworkReply::NetworkError)` is emitted when there is an error with the error code as argument. If you need a user-friendly string, you can retrieve that with `QNetworkReply::errorString()`. `finished()` is emitted when the request is finished. The reply is still open, so you can read it here: `readyRead()` .Since the reply is derived from `QIODevice`, it has the `readyRead` signal, which is emitted whenever more data is ready to read.

On large downloads, you might want to monitor the progress of the download, which is a common thing to do. Usually, every download has some kind of progress bar. `QNetworkReply` emits the `downloadProgress(qint64 bytesReceived, qint64 bytesTotal)` signal as follows:

```
connect(networkReply, &QNetworkReply::downloadProgress, this,
&MyClass::onDownloadProgress);
```

There is the corresponding `uploadProgress` for uploads.

`preSharedKeyAuthenticationRequired(QSslPreSharedKeyAuthenticator *authenticator)` gets emitted when the download needs authentication. The `QSslPreSharedKeyAuthenticator` object should be loaded with the pre-shader key and other details to authenticate the user.

The `sslErrors(const QList<QSslError> &errors)` signal is called when **Secure Sockets Layer (SSL)** encounters problems, including certificate verification errors.

`QNetworkManager` can also do simple **File Transfer Protocol (FTP)** transfers.

QFtp

There are two ways to use FTP with Qt. `QNetworkAccessManager` has simple FTP `get` and `put` support, we can easily use that.

FTP servers usually require some sort of username and password. We use `setUserName()` and `setPassword()` of `QUrl` to set these, as follows:

```
QUrl url("ftp://llornkcor.com/");
url.setUserName("guest@llornkcor.com");
url.setPassword("handsonmobileandembedded");
```

 The source code can be found on the Git repository under the `Chapter05-5` directory, in the `cp5` branch.

Once we know the file's name, we need to add that to the `url`, as it will use this to write the fail, as follows:

```
url.setPath(QFileInfo(file).fileName());
```

Then, set the request `url`, as follows:

```
request.setUrl(url);
```

We can hook up slots to the `QNetworkReply` signals, once we call `put` on the QNAM, as follows:

```
QNetworkReply *networkReply = manager->put(request, fileBytes);

connect(networkReply, &QNetworkReply::downloadProgress,
    this, &MainWindow::onDownloadProgress);
connect(networkReply, &QNetworkReply::downloadProgress,
    this, &MainWindow::onUploadProgress);
```

Do not forget that `error` signal needs `QOverload` as follows:

```
connect(networkReply,
QOverload<QNetworkReply::NetworkError>::of(&QNetworkReply::error),
[=](QNetworkReply::NetworkError code){
    qDebug() << Q_FUNC_INFO << code << networkReply->errorString(); });

connect(networkReply, &QNetworkReply::finished,
    this, &MainWindow::requestFinished);
```

If you need to do more complicated things other than `get` and `put`, you will need to use something else besides `QNetworkAccessManager`.

`QFtp` is not included with Qt, but you can access the standalone `QFtp` module that was ported from Qt 4 to run with Qt 5 as follows:

```
git clone -b 5.12 git://code.qt.io/qt/qtftp.git
```

We will need to build `QFtp`, so we can open the `qtftp.pro` in Qt Creator. Run **Build** and install that.

Using the command line the commands would be as follows:

```
cd qtftp
qmake
make
make install
```

We will need to install this into Qt 5.12, so in Qt Creator, navigate to **Projects** | **Build** | **Build Steps** and select **Add Build Step** | **Make**. In the **arguments** field, type `install`.

Build this and it will also install.

In the project's `.pro` file, to tell `qmake` to use the `network` and `ftp` modules, add the following:

```
QT += network ftp
```

`QFtp` works very typically; log in, do operations, and then log out, as follows:

```
connect(ftp, SIGNAL(commandFinished(int,bool)),
        this, SLOT(qftpCommandFinished(int,bool)));

connect(ftp, SIGNAL(stateChanged(int)),
        this, SLOT(stateChanged(int)));

connect(ftp, SIGNAL(dataTransferProgress(qint64,qint64)),
        this, SLOT(qftpDataTransferProgress(qint64,qint64)));

QUrl url(URL);
ftp->connectToHost(url.host(), 21);
ftp->login(USER, PASS);
```

We connect to the `commandFinished` signal, which can tell us whether there was an error.

The `stateChanged` signal will tell us when we are logged in and the `dataTransferProgress` signal will tell us when bytes are being transferred.

`QFtp` supports other operations, including the following:

- `list`
- `cd`
- `remove`
- `mkdir`
- `rmdir`
- `rename`

QNAM also touches upon my favorite part of Qt Network—Bearer Management. Let's move on to learning about Bearer Management.

Bearer Management of good news

Bearer Management was meant to facilitate user control over the network connections. There are `open` and `close` functions for found connections. One thing it does not do is actually configure these connections. They must already be configured by the system.

It is also meant to be able to group connections together to make it easier to smoothly switch between connections, such as migrating from Wi-Fi to mobile cellular data, somewhat like **Media Independent Handover** (**MIH**) or also **Unlicensed Mobile Access** (**UMA**) specification. If you are interested in an open source library to help with handovers, look at Open MIH at SourceForge.

At the time Qt's Bearer Management was first developed, Symbian was the most used and arguably the most important mobile OS. Symbian had the ability to seamlessly migrate connections between technologies without dropping the connection or data, kind of like the way mobile phone connections get migrated from cell tower to cell tower.

Apple seems to call this Wi-Fi Assist; Samsung has Auto Network Switching.

Years ago, mobile data connections were very expensive, so the connection was often closed after a specific upload or download happened. The opening and closing of connections was more dynamic and needed automatic controls.

At any rate, `QtConfigurationManager` will use what the system supports; it does not implement its own connection data migration.

Qt has the following three main classes that make up Bearer Management:

- `QNetworkConfiguration`
- `QNetworkConfigurationManager`
- `QNetworkSession`

There is also `QBearerEngine`, which is the base class for bearer plugins.

QNetworkConfiguration

`QNetworkConfiguration` represents a network connection configuration, such as a Wi-Fi connection to a particular access point with its **Service Set Identifier** (**SSID**) as the configuration name.

The network configuration can be one of the following types:

- `QNetworkConfiguration::InternetAccessPoint`:
  - This type is a typical access point, such as a Wi-Fi **Access Point** (**AP**) or it could represent an Ethernet or mobile network.

- `QNetworkConfiguration::ServiceNetwork`:
  - A `ServiceNetwork` type is a group of access points known as a **Service Network Access Point** (**SNAP**). The system will determine which of the service networks is best to connect with based on criteria such as cost, speed, and availability. A configuration of the `QNetworkConfiguration::ServiceNetwork` type may also roam between its children `QNetworkConfiguration::InternetAccessPoint`.

- `QNetworkConfiguration::UserChoice`:
  - This type can represent a user preferred configuration. It was used by Nokia's Maemo and Symbian platforms in which the system could pop up a dialog asking the user to choose which AP was best. None of the current bearer backends use this type of `QNetworkConfiguration`.

Often, we need to know the type of bearer, which is to say, what communication protocol the connection is using. Let's find out about `BearerType`.

QNetworkConfiguration::BearerType

This is an `enum` that specifies what the underlying technology of `QNetworkConfiguration` is and can be one of the following:

- `BearerEthernet`
- `BearerWLAN`
- `Bearer2G`
- `BearerCDMA2000`
- `BearerWCDMA`
- `BearerHSPA`
- `BearerBluetooth`
- `BearerWiMAX`
- `BearerEVDO`
- `BearerLTE`
- `Bearer3G`
- `Bearer4G`

This can be determined by calling the `bearerType()` function of the `QNetworkConfiguration` object, as follows:

```
QNetworkConfiguration config;
if (config.bearerType() == QNetworkConfiguration::Bearer4G)
    qWarning() << "Config is using 4G";
```

You can open or connect.

QNetworkConfiguration::StateFlags

`StateFlags` are an OR'd `||`,combination of the `StateFlag` values, which are as follows:

- `Defined`: Known to the system but not yet configured
- `Discovered`: Known and configured, can be used to `open()`
- `Active`: Currently online

A QNetworkConfiguration that has an Active flag will also have the Discovered and Defined flags as well. You can check to see whether a configuration is active by doing this:

```
QNetworkConfiguration config;
if (config.testFlag(QNetworkConfiguration::Active))
    qWarning() << "Config is active";
```

QNetworkConfigurationManager

QNetworkConfigurationManager allows you to obtain QNetworkConfigurations of the system, as follows:

```
QNetworkConfigurationManager manager;
QNetworkConfiguration default = manager.defaultConfiguration();
```

It's always wise to wait for the updateCompleted signal from QNetworkConfigurationManager before using it, to be sure the configurations are set up properly.

A default configuration is the configuration that the system defines as the default. It could have a state of Active or just Discovered.

If you need to simply determine whether the system is currently online, manager->isOnline(); will return true if the system is considered online. Online is when it is connected to another device via a network, which may or may nor be the internet, and may or may not be routed correctly. So, it could be online, but cannot access the internet.

You may need to call updateConfigurations(), which will ask the system to update the list of configurations, and then you need to listen for the updateCompleted signal before proceeding.

You can get all configurations known to the system with a call to allConfigurations(), or filter it to the ones that have a certain state with allConfigurations(QNetworkConfiguration::Discovered);.

In this case, it returns a list of Discovered configurations.

You can check the system's capabilities with a call to `capabilities()`, which can be one of the following:

- `CanStartAndStopInterfaces`: System allows user to start and stop connections
- `DirectConnectionRouting`: Connection routing is bound directly to a specified device interface
- `SystemSessionSupport`: System keeps connection open until all sessions are closed
- `ApplicationLevelRoaming`: Apps can control roaming/migrating
- `ForcedRoaming`: System will reconnect when roaming/migrating
- `DataStatics`: System provides information about transmitted and received data
- `NetworkSessionRequired`: System requires a session

QNetworkSession

`QNetworkSession` provides a way to start and stop connections as well as providing management of connection sessions. In the case of instantiating `QNetworkSession` with a `QNetworkConfiguration` that is a `ServiceNetwork` type, it can provide roaming features. On most systems, roaming will entail actually disconnecting and then connecting a new interface and/or connection. On others, roaming can be seamless and without disturbing the user's data stream.

If the capabilities of `QNetworkConfigurationManager` reports that it supports `CanStartAndStopInterfaces`, then you use `QNetworkSession` to `open()` (connect) and `stop()` (close) `QNetworkConfigurations`.

The QNAM will use `QNetworkSession` when making network requests behind the scenes. You can use `QNetworkSession` to monitor the connection as follows:

 The source code can be found on the Git repository under the `Chapter05-3` directory, in the `cp5` branch.

```
QNetworkAccessManager manager;
QNetworkConfiguration config = manager.configuration();
QNetworkSession *networkSession = new QNetworkSession(config, this);
connect(networkSession, &QNetworkSession::opened, this,
&SomeClass::sessionOpened);
networkSession->open();
```

To monitor bytes received and sent from a QNAM request, connect up to the
`bytesReceived` and `bytesWritten` signals, as follows:

```
connect(networkSession, &QNetworkSession::bytesReceived, this,
&SomeClass::bytesReceived);
connect(networkSession, &QNetworkSession::bytesWritten, this,
&SomeClass::bytesWritten);

QNetworkRequest request(QUrl("http://example.com"));
manager->get(request);
```

Roaming

By roaming, I mean roaming between Wi-Fi and mobile data, not roaming as in out of the
home network, which can be very expensive mobile data to use.

In order to facilitate roaming, a client app can connect to the
`preferredConfigurationChanged` signal and then begin the process by calling
`migrate()` or cancel it by calling `ignore()`. Migrating a connection could be as simple as
pausing the download, disconnecting and reconnecting to the new connection, and then
resuming the download. This method is called forced roaming. It can, on some platforms,
seamlessly migrate the data stream to the new connection, similar to what a mobile phone
does when a call gets migrated to another cell tower.

At this time, there are no currently supported backends that support migrating sessions. A
system integrator could implement a backend that does true connection migration and
handovers. It would also help if the system allows this.

That said, both Samsung's Android and iOS support roaming features seem to have caught
up to where Nokia was years ago. Samsung calls it Adaptive Wi-Fi, previously known as
Smart Network Switch. iOS calls it Wi-Fi Assist. These happen at the system level and
allow roaming between Wi-Fi and mobile data connections. Neither of these platforms
allows applications to control the handover.

QBearerEngine

Qt comes with the following bearer backend plugins based off of the `QBearerEngine` class:

- `Android`: Android
- `Connman`: Linux desktop & embedded, SailfishOS
- `Corewlan`: Mac OS and iOS

- `Generic`: All
- `NativeWifi`: Windows
- `NetworkManager`: Linux
- `NLA`: Windows

Depending on the platform, some of these work in conjunction with the generic backend.

Low level – of sockets and servers

`QTcpSocket` and `QTcpServer` are two classes for sockets used in Qt. They work in much the same way as your web browser and a WWW server. These connect to a network address host, whereas `QLocalSocket` and `QLocalServer` connect to a local file descriptor.

Let's look at `QLocalServer` and `QLocalSocket` first.

In socket server programming, the basic procedure is as follows:

1. Create a socket
2. Set socket options
3. Bind a socket address
4. Listen for connections
5. Accept new connection

Qt simplifies these steps to the following:

1. Create a socket
2. Listen for connections
3. Accept new connection

QLocalServer

If you need communication on the same machine, then QLocalServer will be slightly more performant than using a TCP-based socket server. It can be used for **Inter-process communication (IPC)**.

First, we create the server, and then call the listen function with a string name that clients use to connect. We hook up to the newConnection signal, so we know when a new client connects.

 The source code can be found on the Git repository under the Chapter05-5a directory, in the cp5 branch.

When a client tries to connect, we then send a small message using the write function, and finally flush the message, as follows:

```
QLocalServer *localServer = new QLocalServer(this);
localServer->listen("localSocketName");

connect(localServer, &QLocalServer::newConnection, this,
    &SomeClass::newLocalConnection);

void SomeClass::newLocalConnection()
{
    QLocalSocket *local = localServer->nextPendingConnection();
    local->write("Client OK\r\n");
    local->flush();
}
```

It's that simple! Anytime you need to write to the client, simply use nextPendingConnection() to get the next QLocalSocket object and use write to send the data. Be sure to add \r\n to all lines you need to send, including the last line. The call to flush() is not required, but it sends the data immediately.

You can keep this object around to send more messages when needed.

Our app is now waiting and listening for connections. Let's do that next.

QLocalSocket

QLocalSocket is used to communicate with QLocalServer. You will want to connect to the readyRead signal. Other signals are connected(), disconnected(), error(...), and stateChanged(...), as follows:

> The source code can be found on the Git repository under the Chapter05-5b directory, in the cp5 branch.

```
QLocalSocket *lSocket = new QLocalSocket(this);
connect(lSocket, &QLocalSocket::connected, this, &SomeClass::connected);

connect(lSocket, &QLocalSocket::disconnected, this,
    &SomeClass::disconnected);

connect(lSocket, &QLocalSocket::error, this, &SomeClass::error);
connect(lSocket, &QLocalSocket::readyRead, this, &SomeClass::readMessage);

void SomeClass::readMessage()
{
    if (lSocket->bytesAvailable())
        QByteArray msg = lSocket->readAll();
}
```

If you need state changes, you connect to stateChanged and will be notified when the following states change:

- UnconnectedState
- ConnectingState
- ConnectedState
- ClosingState

Now, we need to actually connect to the server, as follows:

```
lSocket->connectToHost("localSocketName");
```

Like QLocalServer, QLocalSocket uses the write function to send messages to the server, as follows:

```
lSocket->write("local socket OK\r\n");
```

Remember to add the **End Of Line (EOL)** \r\n to mark the end of the data feed line.

That is a simple local sockets based communication. Now, let's look at a TCP-based socket over a network.

QTcpServer

The QTcpServer API is much like QLocalServer and can be pretty much a drop-in replacement with a few small changes. Most notably, the arguments for the listen call are slightly different, in which you need to specify QHostAddress for QTcpServer instead of a QString name, and a port number. Here, I use QHostAddress::Any, which means it will listen on all network interfaces. If you don't care about which port is used, set it to 0 as follows:

```
QTcpServer *tcpServer = new QTcpServer(this);
tcpServer->listen(QHostAddress::Any, 8888);

connect(tcpServer, &QTcpServer::newConnection, this,
    &SomeClass::newLocalConnection);

void SomeClass::newLocalConnection()
{
    QTcpSocket *tSocket = tcpServer->nextPendingConnection();
    tSocket->write("Client OK\r\n");
    tSocket->flush();
}
```

Does it look familiar? QHostAddress can be an IPv4 or IPv6 address. You can also specify different ranges of address by using the QHostAddress::SpecialAddress enum as I did, which can be one of the following:

- LocalHost: 127.0.0.1
- LocalHostIPv6: ::1
- Broadcast: 255.255.255.255
- AnyIPv4: 0.0.0.0
- AnyIPv6: ::
- Any: all IPv4 and IPv6 addresses

QTcpServer has an additional signal to QLocalServer—acceptError, which gets emitted when an error occurs during the accept phase of a new connection. You can also pauseAccepting() and resumeAccepting() the accepting of the connections in the pending connection queue.

QTcpSocket

`QTcpSocket` is similar to `QLocalSocket` as well. Except, among other things, `QTcpSocket` has `connectToHost` as a way to connect to a server, as follows:

```
QTcpSocket *tSocket = new QTcpSocket(this);
connect(tSocket, &QTcpSocket::connected, this, &SomeClass::connected);

connect(tSocket, &QTcpSocket::disconnected, this,
    &SomeClass::disconnected);

connect(tSocket, &QTcpSocket::error, this, &SomeClass::error);
connect(tSocket, &QTcpSocket::readyRead, this, &SomeClass::readData);
```

To make a simple HTTP request, we can write to the socket after we are connected, as follows:

```
void SomeClass:connected()
{
    QString requestLine = QStringLiteral("GET \index.html HTTP/1.1\r\nhost:
www.example.com\r\n\r\n");
    QByteArray ba;
    ba.append(requestLine);
    tSocket->write(ba);
    tSocket->flush();
}
```

This will request the `index.html` file from the server. The data can be read in the `readyRead` signal handler, as follows:

```
void SomeClass::readData()
{
    if (tSocket->bytesAvailable())
        QByteArray msg = tSocket->readAll();
}
```

You can also use the `waitForConnected`, `waitForBytesWritten`, and `waitForReadyRead` functions if you do not want to use this more synchronously, as follows:

```
QTcpSocket *tSocket = new QTcpSocket(this);
if (!tSocket->waitForConnected(3000)) {
    qWarning() << "Not connected";
    return;
}

tSocket->write("GET \index.html HTTP/1.1\r\nhost:
```

```
www.example.com\r\n\r\n");
tSocket->waitForBytesWritten(1000);
tSocket->waitForReadyRead(3000);
if (tSocket->bytesAvailable())
    QByteArray msg = tSocket->readAll();
```

Then, close the connection with the following command:

```
tSocket->close();
```

QSctpServer

SCTP stands for **Stream Control Transmission Protocol**. QSctpServer sends messages as groups of bytes like UDP, rather than a stream of bytes like a TCP socket. It also ensures reliable delivery of the packets, like TCP. It can send several messages in parallel or at the same time. It does this by using several connections.

QSctpServer can also send a stream of bytes like TCP by setting setMaximumChannelCount to −1. The first thing you want to do after creating the QSctpServer object is setMaximumChannelCount. Setting this to 0 will let this use the number of channels that the client uses, as follows:

```
QSctpServer *sctpServer = new QSctpServer(this);
sctpServer->setMaximumChannelCount(8);
```

If you intend to use TCP byte streams, you use the nextPendingConnection() function like QTcpServer to get a QTcpSocket object to communicate with. QSctpServer has the additional nextPendingDatagramConnection() to communicate with QSctpSocket.

To receive bytes in the newConnection signal handler, use the following code:

```
QSctpSocket *sSocket = sctpServer->nextPendingDatagramConnection();
```

QSctpSocket

QSctpSocket also has controls for channel count, and as with QSctpServer, if you set the maximum channel count to −1, it will behave more like TCP sockets and send a data stream instead of message packets. The message blob is called a datagram.

To read and write these datagrams, use readDatagram() and writeDatagram(). Let's examine QNetworkDatagram.

To construct `QNetworkDatagram`, you need a `QByteArray` that holds the data message, a `QHostAddress` for the destination, and optionally, a port number. It can be as simple as the following:

```
QNetworkDatagram datagram("Hello Mobile!", QHostAddress("10.0.0.50"),
8888);
sSocket->writeDatagram(datagram);
```

This will send the `"Hello Mobile!"` message to the corresponding server.

QUdpSocket

`QUdpSocket` sends datagrams such as `QSctpSocket`, but they are not reliable, which means it will not retry to send any datagrams. It is also connectionless and has a restriction on data length of 65,536 bytes.

There are two ways to set up `QUdpSocket`—`bind(...)` and `connectToHost(...)`.

If you use `connectToHost`, you can use `QIODevice read()`, `write()`, `readAll()` to send and receive datagrams. Using the `bind(...)` method, you need to use `readDatagram` and `writeDatagram` instead, as follows:

```
QUdpSocket *uSocket = new QUdpSocket(this);
uSocket->bind(QHostAddress::LocalHost, 8888);
connect(uSocket, &QUdpSocket::readyRead, this, &SomeClass::readMessage);

void SomeClass::readMessage()
{
   while (udpSocket->hasPendingDatagrams()) {
        QNetworkDatagram datagram = uSocket->receiveDatagram();
        qWarning() << datagram.data();
    }
}
```

QSslSocket

Encrypted socket communications can be handled by `QSslSocket`, which uses SSL to encrypt the TCP connection. The encrypted signal is emitted when the connection is secured, as follows:

```
QSslSocket *sslSocket = new QSslSocket(this);
connect(sslSocket, &QSslSocket::encrypted, this,
SomeClass::socketEncrypted);
```

```
sslSocket->connectToHostEncrypted("example.com", 943);
```

 The source code can be found on the Git repository under the
Chapter05-6a directory, in the cp5 branch.

This will start the connection and immediately start the secure handshake procedure. Once
the handshake is finished with no error, the encrypted signal will be emitted and the
connection will be ready.

You will need to add key/certificate pair to QSslSocket to utilize the encryption
capabilities. You can easily generate key-certificate fail pair for testing by using this web
site: https://www.selfsignedcertificate.com/.

Because we are using a self-signed certificate, we will need to add ignoreSslErrors in
our error handling slot:

```
sslSocket->ignoreSslErrors();
```

To add the encryption key and certificate, you need to open and read both files, and use the
resulting QByteArrays to create QSslKey and QSslCertificate:

```
void MainWindow::initCerts()
{
    QByteArray key;
    QByteArray cert;

    QString keyPath =
QFileDialog::getOpenFileName(0, tr("Open Key File"),
                            QDir::homePath(),
                            "Key file (*.key)");

    if (!keyPath.isEmpty()) {
        QFile keyFile(keyPath);
        if (keyFile.open(QIODevice::ReadOnly)) {
            key = keyFile.readAll();
            keyFile.close();
        }
    }

    QString certPath =
QFileDialog::getOpenFileName(0, tr("Open cert File"),
                            QDir::homePath(),
                            "Cert file (*.cert)");

    if (!certPath.isEmpty()) {
```

```
        QFile certFile(certPath);
        if (certFile.open(QIODevice::ReadOnly)) {
            cert = certFile.readAll();
            certFile.close();
        }
    }

    QSslKey sslKey(key, QSsl::Rsa,
QSsl::Pem,QSsl::PrivateKey,"localhost");
    sslSocket->setPrivateKey(sslKey);

    QSslCertificate sslCert(cert);
    sslSocket->addCaCertificate(sslCert);
    sslSocket->setLocalCertificate(sslCert);
}
```

When you run this code, you will need to use `QFileDialog` to navigate and find the `localhost.key` and `localhost.cert` files in the source directory.

Then, we use `setPrivateKey` to set the key file, and `addCaCertificate` and `setLocalCertificate` to add the certificate.

To read from the socket, you can connect to the `readReady` signal like in `QTcpSocket`.

To write to the socket, which transmits to the server, simply use the `write` function:

```
sslSocket->write(ui->lineEdit->text().toUtf8() +"\r\n");
```

You can then use `QSslSocket` to connect to `QTcpServer` that opens `QSslSocket`. This brings us to our next step.

QSslServer

Ok, there is no `QSslServer` class, but since the `QSslSocket` class is just derived from `QTcpSocket` with some extra SSL stuff on top, you can create your own SSL server using the functions from `QSslSocket`.

You will need to generate SSL key and certificates. If they are self-signed, the same rules apply, in which we need to set the following:

```
server->ignoreSslErrors()
```

You can create an SSL server by subclassing `QTcpServer` and overriding `incomingConnection()` ,as follows.

 The source code can be found on the Git repository under the `Chapter05-6` directory, in the `cp5` branch.

We implement the `header` file with the `override` function, as well as a slot to connect to when the server changes into encrypted mode:

```
class MySslServer : public QTcpServer
{
public:
    MySslServer();
protected:
    void incomingConnection(qintptr handle) override;
private slots:
    void socketEncrypted();
};
```

In the implementation of the SSL server class, pay attention to the call to `startServerEncryption()`. This will initiate the encryption of the `server` channels and create a `Server`, as follows:

```
MySslServer::MySslServer()
{
    server = new QSslSocket(this);
    initCerts();
}
```

We also need to add the encruption key and certificate, as this uses `QSslSocket` like in the last section, *QSslSocket*:

```
void MySslServer::incomingConnection(qintptr sd)
{
  if (server->setSocketDescriptor(sd)) {
  addPendingConnection(server);
  connect(server, &QSslSocket::encrypted, this,
&MySslServer::socketEncrypted);
  server->startServerEncryption();
  } else {
  delete server;
  }
}

void MySslServer::socketEncrypted()
```

```
{
  // entered encrypted mode, time to write secure transmissions
}
```

Here, we connect to the `QSslSocket` encrypted signal, which signals when `QSslSocket` enters encrypted mode. From then on, all bytes sent or received are encrypted.

Errors are handled by connecting to the `sslErrors(const QList<QSslError> &errors)` signal:

```
connect(server, QOverload<const QList<QSslError>
&>::of(&QSslSocket::sslErrors),
            [=](const QList<QSslError> &errors){
        for (QSslError error : errors) {
            emit messageOutput(error.errorString());
        }
    });
```

We also need to connect to the `QAbstractSocket::socketError` signal to handle those errors as well:

```
connect(server, SIGNAL(error(QAbstractSocket::SocketError)),
SLOT(error(QAbstractSocket::SocketError)));
```

Other signals you will also want to connect to are the following:

- `QSslSocket::connected`
- `QSslSocket::disconnected`
- `QSslSocket::encrypted`
- `QSslSocket::modeChanged`
- `QSslSocket::stateChanged`

Up till now, we have been using local IP addresses, but what happens when the server is remote and we need not just the server name, but it's IP address? Let's explore how we can use Qt to do domain name lookups.

Lookups – look me up

Computer networks such as the internet rely on **Domain Name Service** (**DNS**) lookups. This is usually done on remote central servers, but can also be used locally.

There are two classes for doing network lookups—QDnsLookup and QHostInfo. QHostInfo will provide simple IP address lookups for a hostname. It is really just looking up an IP address using a hostname. Let's look at how we can use this.

QHostInfo

QHostInfo is a simple class to do address lookups provided by the platform system. It has synchronous, blocking method for lookup, or you can use signal/slots, as follows:

```
QHostInfo hInfo = QHostInfo::fromName("www.packtpub.com");
```

This method blocks until a response is received.

The lookupHost function does asynchronous lookups and takes a slot as an argument, as follows:

```
QHostInfo::lookupHost("www.packtpub.com", this,
SLOT(lookupResult(QHostInfo)));
```

The slot we need to implement receives QHostInfo as an argument, as such:

```
void SomeClass::lookupResult(QHostInfo info)
{
    if (!hInfo.addresses().isEmpty()) {
        QHostAddress address = info.addresses().first();
        qWarning() << address.toString();
    }
}
```

To get an address from either of these responses, do something like the following:

```
if (!hInfo.addresses().isEmpty()) {
    QHostAddress address = info.addresses().first();
    qWarning() << address.toString();
}
```

Let's now proceed to QDnsLookup.

QDnsLookup

QDnsLookup can look up different types of records, not just IP addresses. The values you can use to set the type of lookup are as follows:

- A: IPv4 addresses, access with hostAddressRecords()
- AAAA: IPv6 addresses, access with hostAddressRecords()
- ANY: Any record
- CNAME: Canonical name, access with canonicalNameRecords()
- MX: Mail exchange, access with mailExchangeRecords()
- NS: Name server, access with nameServerRecords()
- PTR: Pointer, access with pointerRecords()
- SRV: Service, access with serviceRecords()
- TXT: Text, access with textRecords()

Let's look at how this can be implemented. We connect the QDnsLookup signal named finished to our lookupFinished slot. We set the type here to TXT to access text records:

```
QDnsLookup *lookup = new QDnsLookup(this);
connect(lookup, &QDnsLookup::finished, this, &SomeClass::lookupFinished);
lookup->setType(QDnsLookup::TXT);
lookup->setName("example.com");
lookup->lookup();
```

The call to lookup() will start a lookup of the text records for the name that we set, which is example.com. We still need to handle the response, as follows:

```
void SomeClass:: lookupFinished()
{
    QDnsLookup *lookup = qobject_cast<QDnsLookup *>(sender());
    if (!lookup)
        return;
    if (lookup->error() != QDnsLookup::NoError) {
        lookup->deleteLater();
        return;
    }
    const QList<QDnsTextRecord> txtRecords = lookup->textRecords();
    for (const QDnsTextRecord &record: txtRecords) {
        const QString recordName = record->name();
        const QList <QByteArray> recordValues = record->values();
```

```
          . . .
      }
  }
```

You can then use these records in the manner you need.

Summary

QNetwork is quite extensive in what it can do. I have touched upon a few features, such as QNetworkRequest, QNetworkAccessManager, and QNetworkReply to make network requests, such as get and put. You can use Qt's Bearer Management features to control the online state and QNetworkSession to group connections together to roam between connections. We discussed socket development with QLocalSocket, QLocalServer, QTcpSocket, and QTcpServer. You can perform host and DNS lookups with QHostInfo and QDnsLookup.

Connectivity can mean a few things, and in the next chapter, we explore connectivity using Bluetooth **Low Energy** (**LE**).

Connectivity with Qt Bluetooth LE

6

You will learn about using Qt Bluetooth **Low Energy** (**LE**) to build connectivity to devices that have LE Bluetooth radios. Bluetooth is more than mice, keyboards, and audio. Device discovery, data exchange, and other tasks involving Bluetooth Low Energy will be examined. We will use the `QBluetoothUuid`, `QBluetoothCharacteristic`, `QLowEnergyController`, and `QLowEnergyService` classes.

We will cover the following topics in this chapter:

- What is Bluetooth Low Energy
- Discovering and connecting with devices
- Advertising services
- Retrieving sensor data from remote device

What is Bluetooth Low Energy?

Bluetooth Low Energy (**BLE**), or Bluetooth Smart as it is also called, was originally developed by Nokia under the name Wibree and was originally released in 2006. It was integrated into the Bluetooth 4.0 specification and released in 2010.

Bluetooth is a wireless connection technology that operates in the 2,400-2,483.5 MHz range of the 2.4 GHz frequency band. There are 79 data channels it can choose for transmitting packets. BLE limits the data channels to 40.

BLE is targeted at mobile and embedded devices that require lower power consumption. Unlike Bluetooth, BLE is designed for devices that exchange small amounts of data periodically, as opposed to regular Bluetooth that was designed for continuous data streams. Most importantly, BLE has a sleep mode that it uses to conserve power.

Qt has support for BLE in the Qt Connectivity module alongside **Near-field Communication** (NFC). BLE has a number of profiles and services:

- Alerts
- Battery
- Fitness
- Health
- HID
- Internet
- Mesh
- Sensors

Generic Attribute (**GATT**) is used to store profiles, services, characteristics, and other data. Each entry is a unique 16-bit ID. The BLE connection is exclusive in that it can only connect to one computer at a time. The BLE peripheral device is known as the GATT server, and the computer it connects to is the GATT client.

Each profile can have a number of services. Each service can have a number of characteristics. A profile is just the collection of pre-defined services in the specification.

A service is just a group of characteristics defined by a unique 16 or 128-bit ID. A characteristic is a single data point, which may contain an array of data, such as with an accelerometer.

Now that you know a little bit of the background, let's get started.

Implementing a BLE GATT server

I guess we really need a BLE server now.

Let's say you have an embedded device that has a few environmental sensors attached, such as humidity and temperature. You need to send this data over Bluetooth to another handheld device once in a while. On the embedded sensor device, you need to setup the device. The basic procedure to set up the BLE server is as follows:

1. Supply advertisement data (QLowEnergyAdvertisingData)
2. Supply characteristic data (QLowEnergyCharacteristicData)
3. Set up the service data (QLowEnergyServiceData)
4. Start advertising and listening for connections

QLowEnergyAdvertisingData

`QLowEnergyAdvertisingData` is the class you use to tell the server what and how the data is going to be presented.

Here's how we would use `QLowEnergyAdvertisingData`.

Construct a `QLowEnergyAdvertisingData` object:

```
QLowEnergyAdvertisingData *leAdd = new QLowEnergyAdvertisingData;
```

Set `Discoverability` options:

```
leAdd->setDiscoverability(
QLowEnergyAdvertisingData::DiscoverabilityGeneral);
```

Set a `Name` for our service:

```
leAdd->setLocalName("SensorServer");
```

Add a list of services we are interested it:

```
QList<QBluetoothUuid> servicesList
<< QBluetoothUuid::EnvironmentalSensing;
leAdd->setServices(servicesList);
```

 The source code can be found on the Git repository under the `Chapter06-1` directory, in the `cp6` branch.

We need to create some characteristic data now. Let's create a `Characteristic` that handles temperature, so we set its `uuid` to `TemperatureMearurement`. We need to also let it be configurable for notifications.

QLowEnergyCharacteristicData

`QLowEnergyCharacteristicData` represents a **Generic Attribute Profile (GATT)** characteristic, which defines a single data point in the Bluetooth transfer. You use it to set up service data:

```
QLowEnergyCharacteristicData chData;
chData.setUuid(QBluetoothUuid::TemperatureMeasurement);
```

```
chData.setValue(QByteArray(2,0));
chData.setProperties(QLowEnergyCharacteristic::Notify);
const QLowEnergyDescriptorData
descriptorData(QBluetoothUuid::ClientCharacteristicConfiguration,
QByteArray(2, 0));
chData.addDescriptor(descriptorData);
```

QLowEnergyServiceData

Here, we set up the `Temperature` service data as a `Primary` service, and add `Characteristic` to service:

```
QLowEnergyServiceData serviceData;
serviceData.setUuid(QBluetoothUuid::Temperature);
serviceData.setType(QLowEnergyServiceData::ServiceTypePrimary);
serviceData.addCharacteristic(chData);
```

Now, let's supply the temperature data. We construct `QLowEnergyCharacteristic` with the `TemperatureMeasurement` type, and supply to it some data. The first bit specifies that we are supplying the `temperature` unit in Celsius:

```
QLowEnergyCharacteristic characteristic =
service->characteristic(QLowEnergyCharacteristic::TemperatureMeasurement);
quint8 temperature = 35;

QByteArray currentTempValue;
value.append(char(0));
value.append(char(temperature));
service->writeCharacteristic(characteristic, currentTempValue);
```

We are all set up now, and all we need is to start `Advertising` to listen for connections:

```
controller->startAdvertising(QLowEnergyAdvertisingParameters(), leAdd,
leAdd);
```

Discovery and Pair-ity – search and connect for BLE devices

The first thing you need to do is search for devices, which is called discovery. It entails putting the Bluetooth device into search, or discovery mode. You then receive a list of devices address with which you can connect or pair to be able to access and share data.

Let's look at how that is done in Qt using `QBluetoothDeviceDiscoveryAgent`.

QBluetoothDeviceDiscoveryAgent

The `QBluetoothDeviceDiscoveryAgent` class is responsible for the device discovery search. It will emit the `deviceDiscovered` signal when any Bluetooth is found:

```
QBluetoothDeviceDiscoveryAgent *discoveryAgent = new
QBluetoothDeviceDiscoveryAgent(this); connect(discoveryAgent,
SIGNAL(deviceDiscovered(QBluetoothDeviceInfo)),
        this, SLOT(newDevice(QBluetoothDeviceInfo)));
discoveryAgent->start(QBluetoothDeviceDiscoveryAgent::LowEnergyMethod));
```

The source code can be found on the Git repository under the `Chapter06-1a` directory, in the `cp6` branch.

The call to `start()` will initiate the discovery process. The `QBluetoothDeviceDiscoveryAgent::LowEnergyMethod` argument will set a filter to only discover `LowEnergy` devices. Once you find the device you want, you can call `stop()` to stop the device search.

You can wait for errors by connecting to the error (`QBluetoothDeviceDiscoveryAgent::Error error`) signal.

The `error` signal in the `QBluetoothDeviceDiscoveryAgent` class is overloaded, so special care needs to happen in order to connect to the signal. Qt provides `QOverload` and can be implemented like this:

```
connect(discoveryAgent,
QOverload<QBluetoothDeviceDiscoveryAgent::Error>::of(&QBluetoothDeviceDisco
veryAgent::error), this, &SomeClass::deviceDiscoveryError);
```

If you would rather get a list of devices all at one time, connect to the `Finished` signal and use the `discoveryDevices()` call, which returns `QList <QBluetoothDeviceInfo>`:

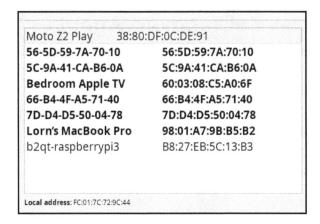

```
Moto Z2 Play        38:80:DF:0C:DE:91
56-5D-59-7A-70-10        56:5D:59:7A:70:10
5C-9A-41-CA-B6-0A        5C:9A:41:CA:B6:0A
Bedroom Apple TV        60:03:08:C5:A0:6F
66-B4-4F-A5-71-40        66:B4:4F:A5:71:40
7D-D4-D5-50-04-78        7D:D4:D5:50:04:78
Lorn's MacBook Pro        98:01:A7:9B:B5:B2
b2qt-raspberrypi3        B8:27:EB:5C:13:B3

Local address: FC:01:7C:72:9C:44
```

You might want to check for the remote devices pairing status, so call `pairingStatus` of `QLocalBluetoothDevice`.

You can pair with a device by then calling the `requestPairing` function of `QBluetoothLocalDevice`, with `QBluetoothAddress` of the remote Bluetooth device:

```
SomeClass::newDevice(const QBluetoothDeviceInfo &info)
{
    QBluetoothLocalDevice::Pairing pairingStatus =
localDevice->pairingStatus(info.address());
    if (pairingStatus == QBluetoothLocalDevice::Unpaired) {
        QMessageBox msgBox;
        msgBox.setText("Bluetooth Pairing.");
        msgBox.setInformativeText("Do you want to pair with device: " +
item->data(Qt::UserRole).toString());
        msgBox.setStandardButtons(QMessageBox::Ok | QMessageBox::Cancel);
        msgBox.setDefaultButton(QMessageBox::Cancel);
        int ret = msgBox.exec();
        if (ret == QMessageBox::Ok) {
            qDebug() << Q_FUNC_INFO << "Pairing...";
            localDevice->requestPairing(address,
QBluetoothLocalDevice::Paired);
    }

    }
}
```

Our example app asks to pair the device before we execute the `requestPairing` procedure:

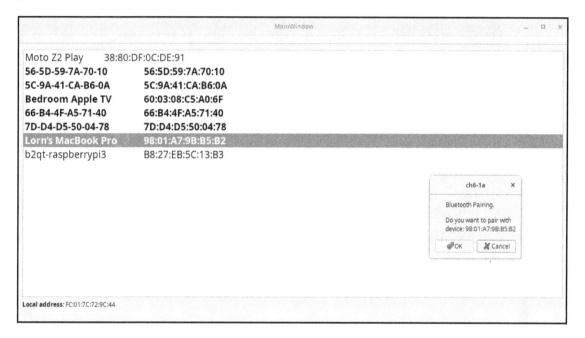

You can then call `requestPairing` on `QBluetoothLocalDevice` with the `QBluetoothAddress` of the device you wish to pair with. Let's take a look at `QBluetoothLocalDevice`

QBluetoothLocalDevice

`QBluetoothLocalDevice` represents the Bluetooth on your device. You use this class to initiate pairing to another device, but also to handle pairing requests from remote Bluetooth devices. It has a few signals to help with that:

- `pairingDisplayConfirmation`: This is a signal the remote device requests to show user a PIN and ask whether it is the same on both devices. You must call `pairingConfirmation` with `true` or `false` on `QBluetoothLocalDevice`.
- `pairingDisplayPinCode`: This is a request to enter a PIB.
- `pairingFinished`: Pairing is completed successfully.

We then connect to these signals, if the user allows it when they click on the **OK** button:

```
        if (ret == QMessageBox::Ok) {

            connect(localDevice,
&QBluetoothLocalDevice::pairingDisplayPinCode, this,
&MainWindow::displayPin);
            connect(localDevice,
&QBluetoothLocalDevice::pairingDisplayConfirmation, this,
&MainWindow::displayConfirmation);
            connect(localDevice, &QBluetoothLocalDevice::pairingFinished,
this, &MainWindow::pairingFinished);
            connect(localDevice, &QBluetoothLocalDevice::error, this,
&MainWindow::pairingError);
            localDevice->requestPairing(address,
QBluetoothLocalDevice::Paired);
        }
```

When the remote device only needs a PIN confirmation, the
`pairingDisplayConfirmation` signal is called:

```
SomeClass::displayConfirmation(const QBluetoothAddress &address, const
QString &pin)
{
    QMessageBox msgBox;
    msgBox.setText("Confirm pin");
    msgBox.setInformativeText("Confirm the pin is the same as on the
device.");
    msgBox.setStandardButtons(QMessageBox::Ok | QMessageBox::Cancel);
    msgBox.setDefaultButton(QMessageBox::Cancel);
    int ret = msgBox.exec();
    if (ret == QMessageBox::Ok) {
        localDevice->pairingConfirmed(true);
      } else {
        localDevice->pairingConfirmed(false);
    }
}
```

When the remote device needs user to enter a PIN, the `pairingDisplayPinCode` signal is
called with a PIN to be displayed and entered on the remote device:

```
SomeClass::displayPin(const QBluetoothAddress &address, const QString &pin)
{
{
    QMessageBox msgBox;
    msgBox.setText(pin);
    msgBox.setInformativeText("Enter pin on remote device");
    msgBox.setStandardButtons(QMessageBox::Ok);
```

```
        msgBox.exec();
    }
```

On the other side, to receive pairing, you need to put `QBluetoothLocalDevice` into the `Discoverable` mode:

```
    localDevice->setHostMode(QBluetoothLocalDevice::HostDiscoverable);
```

The device can then be seen by other devices that are in the Bluetooth `Discovery` mode.

Specifying and getting client data

Once you have connected to a BLE device peripheral, you need to discover its characteristics to be able to read and write them. You do that by using `QLowEnergyController`. Let's take a look at what `QLowEnergyController` is.

QLowEnergyController

`QLowEnergyController` is the central place to access BLE devices both local and remote.

The local `QLowEnergyController` can be created by using the static `QLowEnergyController::createPeripheral(QObject *parent)` function.

Creating a `QLowEnergyController` object that represents the remote device is done by calling the static class `QLowEnergyController::createCentral` using the `QBluetoothDeviceInfo` object that you receive when you discover remote devices.

The `QLowEnergyController` object has several signals:

- `discoveryFinished`
- `serviceDiscovered`
- `connected`
- `disconnected`

Connect to the `connected` signal and start connecting by calling `connectToDevice()`:

```
SomeClass::newDevice(const QBluetoothDeviceInfo &device)
{
    QLowEnergyController *controller = new
QLowEnergyController(device.address());
    connect(controller, &QLowEnergyController::connected, this,
&SomeClass::controllerConnected);

    controller->connectToDevice();
}

SomeClass::controllerConnected()
{
    QLowEnergyController *controller = qobject_cast<QLowEnergyController
*>(sender());
    if (controller) {
        connect(controller, &QLowEnergyController::serviceDiscovered, this,
&SomeClass::newServiceFound);
        controller->discoverServices();
    }
}
```

Once the device is connected, it's time to discover its services, so we connect to the `serviceDiscovered` signal and initiate the service discovery by calling `discoverServices()`.

QLowEnergyService

You can also connect to the `discoveryFinished()` signal, which returns a list of discovered services by calling `services()`. With either of these, you will get the `QBluetoothUuid` that belongs to that service, with which you can then create a `QLowEnergyService` object:

```
SomeClass::newServiceFound(const QBluetoothUuid &gatt)
{
    QLowEnergyController *controller = qobject_cast<QLowEnergyController
*>(sender());
    QLowEnergyService *myLEService = controller->createServiceObject(gatt,
this);
}
```

We now have a `QLowEnergyService` object, which gives us details about it. We can only read its service details when its state becomes `ServiceDiscovered`, so now call the `discoverDetails()` function of the service to start the discovery process:

```
QLowEnergyService *myLEService = controller->createServiceObject(gatt,
this);
    connect(myLEService, &QLowEnergyService::stateChanged, this,
&SomeClass::serviceStateChanged);
    myLEService->discoverDetails();
```

Let's have a look at `QLowEnergyCharacteristic`.

QLowEnergyCharacteristic

Once the service details or `characteristics` are discovered, we can perform actions with `QLowEnergyCharacteristic`, such as enabling notifications:

```
void SomeClass::serviceStateChanged(QLowEnergyService::ServiceState state))
{
    if (state != QLowEnergyService::ServiceDiscovered)
        return;
    QLowEnergyService *myLEService = qobject_cast<QLowEnergyService
*>(sender());
    QList <QLowEnergyCharacteristic> characteristics =
myLEService->characteristics();

}
```

Using `QLowEnergyCharacteristic`, we can get a `QLowEnergyDescriptor` that we use to enable or disable notifications.

Sometimes, a `characteristic` on the remote device needs to be written to as well, such as enabling a specific sensor. In this case, you need to use the `writeCharacteristic` function of the service with `characteristic` as the first argument and the value to be written as the second:

```
QLowEnergyCharacteristic *movementCharacteristic =
myLEService->characteristic(someUuid);
myLEService->writeCharacteristic(movementCharacteristic,
QLowEnergyCharacteristic::Read);
```

Writing to `QLowEnergyDescriptor` is just as easy; let's take a look.

QLowEnergyDescriptor

From the Bluetooth specifications, a descriptor is defined as attributes that describe a characteristic value. It contains additional information about a characteristic. `QLowEnergyDescriptor` encapsulates a GATT descriptor. Descriptors and characteristics can have notifications when changes happen.

To enable notifications, we might need to write a value to the descriptor. Here are some possible values:

GATT term	Description	Value	Qt constant
Broadcast	Permits broadcast	0x01	QLowEnergyCharacteristic::Broadcasting
Read	Permits reading	0x02	QLowEnergyCharacteristic::Read
Write without response	Permits writing with any response	0x04	QLowEnergyCharacteristic::WriteNoResponse
Write	Permits writing with a response	0x08	QLowEnergyCharacteristic::Write
Notify	Permits notifications	0x10	QLowEnergyCharacteristic::Notify
Indicate	Permits notification with client confirmation required	0x20	QLowEnergyCharacteristic::Indicate
Authenticated signed writes	Permits signed writes	0x40	QLowEnergyCharacteristic::WriteSigned
Extended properties	Queued writes and writable auxiliaries	0x80	QLowEnergyCharacteristic::ExtendedProperty

The difference between notifications and indications is that, with indications, the server requires the client to confirm that it has received the message, whereas with a notification, the server doesn't care whether the client receives it.

 Qt does not currently have support to use authenticated signed writes (0x40) with Qt, nor does it have support to use indications (0x20).

We want to be notified when the characteristic values change. To enable this, we need to write a value of `0x10` or `QLowEnergyCharacteristic::Notify` to descriptor:

```
for ( const QLowEnergyCharacteristic character :  characteristics) {
    QLowEnergyDescriptor descriptor =
character.descriptor(QBluetoothUuid::ClientCharacteristicConfiguration);
    connect(myLEService, &QLowEnergyService::characteristicChanged, this,
```

```
&SomeClass::characteristicUpdated);
    myLEService->writeDescriptor(descriptor,
QByteArrayLiteral("\x01\x00"));
}
```

Or we can use the predefined QLowEnergyCharacteristic::Notify, like so:

```
myLEService->writeDescriptor(descriptor,
QLowEnergyCharacteristic::Notify));
```

Now, we can finally get values out of our Bluetooth LE device:

```
void SomeClass::characteristicUpdated(const QLowEnergyCharacteristic &ch,
const QByteArray &value)
{
    qWarning() << ch.name() << "value changed!" << value;
}
```

Bluetooth QML

There are Bluetooth QML components you can use as a client to scan and connect to Bluetooth devices as well. They are simple but functional.

 The source code can be found on the Git repository under the Chapter06-2 directory, in the cp6 branch.

1. Add the bluetooth module to your pro file:

    ```
    QT += bluetooth
    ```

2. In your qml file, use the QtBluetooth import:

    ```
    import QtBluetooth 5.12
    ```

The most important element is BluetoothDiscoveryModel.

BluetoothDiscoveryModel

`BluetoothDiscoveryModel` provides a data model of available Bluetooth devices nearby. You can use it in various model-based Qt Quick components, such as `GridView`, `ListView`, and `PathView`. Setting the `discoveryMode` property tells the local Bluetooth device the level of service discovery, which is one of the following:

- `FullServiceDiscovery`: Discovers all services of all devices
- `MinimalServiceDiscovery`: Minimal discovery only includes device and UUID information
- `DeviceDiscovery`: Discovers only devices and no services

The discovery process will take various amounts of time according to the number of services that need to be discovered. To speed up the discovery of a specific device, you can set the `discoveryMode` property to `BluetoothDiscoveryModel.DeviceDiscovery`, which will allow you to discover the target device address. In the following example, I have commented out the device's target Bluetooth address so it will at least show some devices when you run it:

```
BluetoothDiscoveryModel {
    id: discoveryModel
    discoveryMode: BluetoothDiscoveryModel.DeviceDiscovery
    onDeviceDiscovered: {
        if (/*device == "01:01:01:01:01:01" && */ discoveryMode ==
BluetoothDiscoveryModel.DeviceDiscovery) {
            discoveryModel.running = false
            discoveryModel.discoveryMode =
BluetoothDiscoveryModel.FullServiceDiscovery
            discoveryModel.remoteAddress = device
            discoveryModel.running = true
        }
    }
}
```

To discover all services of all nearby devices, set `discoveryMode` to `BluetoothDiscoveryModel.FullServiceDiscovery`. If you set the `remoteAddress` property with a device address, you can target that one specific device. You will then have to toggle the `running` property off and then on to start a new scan.

We have a basic data model, but we need somewhere to display it. Qt Quick has a few options for viewing model data:

- `GridView`
- `ListView`
- `PathView`

> `PathView` is best written with Qt Creator QML designer, as you can visually adjust its path.

Let's choose a `ListView` for simplicity although I really wanted to use `PathView`:

```
ListView {
    id: mainList
    anchors.top: busy.bottom
    anchors.fill: parent
    model: discoveryModel
}
```

It's not going to show anything without defining `delegate`:

```
delegate: Rectangle {
    id: btDelegate
    width: parent.width
    height: column.height + 10
    focus: true
    Column {
        id: column
        anchors.horizontalCenter: parent.horizontalCenter
        Text {
            id: btText
            text: deviceName ? deviceName : name
            font.pointSize: 14
        }
    }
}
```

Scanning for devices can take a while to complete sometimes, so I want to add a busy indicator. Qt Quick Control 2 has `BusyIndicator`:

```
BusyIndicator {
    id: busy
    width: mainWindow.width *.6
    anchors.horizontalCenter: parent.horizontalCenter
```

```
        anchors.top: mainWindow.top
        height: mainWindow.height / 8
        running: discoveryModel.running
    }
```

When you discover remote services, you will get a `BluetoothService` object.

BluetoothService

When you specify `BluetoothDiscoveryModel.FullServiceDiscovery` for a discovery scan and when `BluetoothDiscoveryModel` locates a new service, the `serviceDiscovered` signal will be emitted. When we connect to that signal, we will receive the `BluetoothService` object in the slot.

We can the get the **universal unique identifier (uuid)**, device and service name, service description, and other details. You can use this `BluetoothService` to connect to `BluetoothSocket`.

BluetoothSocket

The `BluetoothSocket` component can be used to send and receive `String` messages.

To implement this component, at it's simplest would be the following:

```
BluetoothSocket {
    id: btSocket
}
```

 `BluetoothSocket` does not handle binary data. For that, you will have to use the C++ `QBluetoothSocket` class.

In `BluetoothDiscoveryModel`, handle the `serviceDiscovered` signal. You will get a `BluetoothService` object named `service`. You can then set `Socket` to use the service with the `setService` method:

```
onServiceDiscovered {
    if (service.serviceName == "Magical Service")
        btSocket.setService(service)

}
```

First, you might want to handle the `stateChanged` signals:

```
onSocketStateChanged: {
  switch (socketState) {
  case BluetoothSocket.Unconnected:
  case BluetoothSocket.NoServiceSet:
  break;
  case BluetoothSocket.Connected:
  console.log("Connected");
  break;
  case BluetoothSocket.Connecting:
  console.log("Connecting...");
  break;
  case BluetoothSocket.ServiceLookup:
  console.log("Looking up Service");
  break;
  case BluetoothSocket.Closing:
  console.log("Closing connection");
  break;
  case BluetoothSocket.Listening:
  console.log("Listening for incoming connections");
  break;
  case BluetoothSocket.Bound:
  console.log("Bound to local address")
  break;
  }
  }
```

To connect to the service, write `true` to the `connected` property:

```
btSocket.connected = true
```

Once the `socketState` property is `Connected`, you can transmit a message or string data using the `stringData` property:

```
btSocket.stringData = "Message Ok"
```

Qt Quick offers a simple way to send string messages over Bluetooth.

Summary

Bluetooth Low Energy is meant to have lower energy requirements for mobile and embedded devices. Qt offers both C++ and QML classes and components to use it. You should now be able to discover and connect to a Bluetooth Low Energy device.

Advertising GATT services so users and clients can receive and send data was also covered.

In the next chapter, we will go over some of the main components for the **Internet of Things (IoT)**, such as sensors and automation communication protocols.

Machines Talking

7

Machine automation and IoT use various APIs for communication with each other.

I like to say that you cannot have IoT without sensors. They truly define IoT. Sensors are everywhere these days. Cars, lights, and mobile phones all have a myriad of sensors. Laptop computers have led, light, touch, and proximity sensors.

MQTT and WebSockets are communication and messaging protocols. One use of them is to send sensors to remote locations.

You will learn about using Qt APIs for machine-to-machine automation and communication to web applications using the `QWebSocket` and `QWebSocketServer` classes.

MQTT is a publish-and-subscribe-based TCP/IP protocol for sending sensor data over a limited bandwidth network using `QMqttMessage` to a `QMqttClient` and `QMqttSubscription`.

We will be covering the following topics:

- **Sensory control** - QtSensor data
- **WebSockets** - Bi-directional web communication
- **QMqtt** - Brokers of machine talk

Sensory control – QtSensor data

The Qt Sensors API started with Qt Mobility, which itself grew from Qtopia, which was later renamed Qt Extended.

Qt Mobility was a collection of APIs useful for mobile and embedded devices. It was intended specifically for use in Nokia phones. Some of the Mobility API was integrated into Qt 4 and later into Qt 5.

Qt Sensors, on the other hand, was put into its own repository when Qt 5 split into modules. Qt Sensors started out targeting mobile phone platforms, but as computers, such as laptops and Raspberry Pis, gained sensors, the backends expanded. You can find backends for iOS, Android, WinRT, generic Linux, Sensorfw, as well as Texas Instrument's SensorTag. At my GitHub repository, you can find additional sensor backends for Raspberry Pi Sense HAT, and MATRIX Creator for Raspberry Pi.

Sensor Framework (**SensorFW**) is a framework and backend for configuring and reading sensor data in a variety of ways. It is tried, tested, and used on some of the best alternative mobile devices. It has support for Hybris (which is used in Sailfish OS), Linux IIO sensors, as well as for reading directly from the Linux filesystem. If you are integrating a new device and need to read various sensors, I recommend using Sensor Framework, available from `https://git.merproject.org/mer-core/sensorfw/`.

There are dozens of different sensors for monitoring the environment. Qt Sensors handles the most common sensors found in mobile phones and tablets, and provides tools to help implement new sensor types that may be developed and become popular.

Not only are sensors for monitoring the environment; they can also be used as an input to the system. The Qt Sensor API includes an ad-hoc `QSensorGestures`, which is an API for various device gestures, such as shake, free-fall, hover, cover, turnover, and pickup.

Qt Sensors has the C++ and QML APIs. Let's start with the C++ API.

There are actually three ways to use this API. The first is the generic way. All the sensor classes are derived from `QSensor`. A more generic way to use them is to just use `QSensor`.

QSensor

`QSensor` has two static functions that we can use. `QSensor::sensorTypes()` which returns a `QList` of sensor types; for example, it could be `QLightSensor` or `QOrientationSensor`. You can then use `QSensor::sensorForType` or `defaultSensorForType`. Usually there is only one sensor for a type, so, using the latter will suffice.

But first, we need to tell `qmake` that we want to use the `sensors` module, so in the `.pro` file, do the following:

 The source code can be found on the Git repository under the `Chapter07-1` directory, in the `cp7` branch.

```
QT += sensors
```

To include all `QSensors` headers, the include file line is `#include <QtSensors>`, so let's add this to our file.

Get a list of all sensor types known to the system by using `QSensor::sensorTypes()`:

```
for (const QByteArray &type : QSensor::sensorTypes()) {
    const QByteArray &identifier = QSensor::defaultSensorForType(type);
```

`QSensor` is created by supplying a `QSensor::type` argument, and then you call the `setIdentifier` function with a `String` indicating the sensor you want to use.

```
QSensor* sensor = new QSensor(type, this);
sensor->setIdentifier(identifier);
```

We now have a `QSensor`. You must then call `connectToBackend()` if you are using `QSensor` directly:

```
if (!sensor->connectToBackend())
    qWarning() << "Could not connect to sensor backend";
```

You can then connect to the `readingChanged()` signal and read the values from there. To get the `QSensor`, you can use the `sender()` function in any slot, and then `qobject_cast` to cast to a `QSensor`:

```
connect(sensor, &QSensor::readingChanged, this,
&SomeClass::readingChanged);
```

The `readingChanged()` slot looks like this:

```
void SomeClass::readingChanged()
{
    QSensor *sensor = qobject_cast<QSensor *>(sender());
    QSensorReading *reading = sensor->reading()
    QString values;
    for (int i = 0; i < reading->valueCount(); i++) {
        values += QString::number(reading->value(i).toReal()) + " ";
    }
```

```
    ui->textEdit->insertPlainText(sensor->type() +" " + sensor->identifier()
+ " "+ values + "\n");
}
```

We cast the QSensor using the sender() function, which returns the object that the slot is connected to. We then use that to get the QSensorReading using the reading() function. From the reading, we can get the values the sensor signaled to us.

We still need to call start() on the sensor, so we will add this somewhere after we connect to the readingChanged() signal. This will activate the sensor's backend and start reading data from the device.

```
if (!sensor->isActive())
    sensor->start();
```

There is another way to access a sensor, and that is by using a QSensor subclass. Let's have a look at how we will use QSensor as a subclass:

The QSensor subclass

A more popular way to use Qt Sensors is to use the standard QSensors derived classes, such as QLightSensor or QAccelerometer. This is useful if you know exactly which sensors your device has or what you are going to use. It also reduces the need for type-casting. In this way, it is also easier to use a class's sensor-specific properties:

```
QLightSensor *lightSensor = new QLightSensor(this);
if (!lightSensor->connectToBackend()) {
    qWarning() << "Could not connect to light sensor backend";
    return;
}
connect(lightSensor, &QLightSensor::readingChanged,
&SomeClass::lightSensorChanged);
```

Instead of a generic QSensorReading, we get a sensor specific reading, QLightReading in this case, with a sensor-specific value accessor:

```
SomeClass::lightSensorChanged(const QLightReading *reading)
{
    qWarning() << reading->lux();
}
```

Another way to access sensor data is to use a QSensorFilter. Let's go there.

QSensorFilter

There is a third way to access sensor data in C++, which is to use the sensor-specific filter class. This provides an efficient callback when signals and slots might be too slow, as in the case of `QAccelerometer` and other motion sensors which might be running at 200 cycles per second. It also provides a way to apply one or more filters that affect the values before they get emitted by the sensor reading signals. You could provide additional smoothing and noise reduction, or amplify the signal to a greater range.

In our case, our class would inherit from `QLightFilter`.

```
class LightFilter : public QLightFilter
{
public:
```

We then implement the filter override.

If the `filter` function returns `true`, it will store the `QLightReading` of the `QLightSensor` and the new values will be emitted by, in our case, the `QLightSensor` class. Let's apply a simple moving-average filter to our light sensor data:

```
bool filter(QLightReading *reading)
{
    int lux = 0;
      int averageLux = 0;
      if (averagingList.count() <= 4) {
         averagingList.append(reading->lux());
      } else {
         for (int i = 0; i < averagingList.count(); i++) {
            lux += averagingList.at(i);
         }
         averageLux = lux / (averagingList.count());
         reading->setLux(averageLux);
         averagingList.append(averageLux);
         return true; // store the reading in the sensor
      }
      return false;
   };
   QList<int> averagingList;

};
```

You can then create a new `LightFilter` object and set `QLightSensor` to use it. Add this before the call to `start()`:

```
if (type == QByteArray("QLightSensor")) {
    LightFilter *filter = new LightFilter();
    sensor->addFilter(filter);
}
```

Now let's find out about the `QSensor` data and how to access it.

QSensor data

`QSensor` has values that are specific to the respective sensor. You can access them either generically with `QSensor`, or by sensor value.

QSensorReading

If you are using the more generic `QSensor` class, there is a corresponding `QSensorReading` that you can use to retrieve the generic data. For getting any sensor-specific data you will need to use the corresponding sensors' `QSensorReading` subclass, such as `QAccelerometerReading`. For example, if we are using the `QSensor` to grab accelerometer data, we could do the following:

```
QSensorReading reading;
QList <qreal> data;
qreal x = reading.at(0);
qreal y = reading.at(1);
if (reading.valueCount() == 3)
    qreal z = reading.at(2);
qreal timestamp = reading.timestamp;
```

However, using the `QAccelerometer` and `QAccelerometerReading` classes to do the same thing, would look like this.

```
QAccelerometer accel;
QAccelerometerReading accelReading = accel.reading();
qreal x = accelReading.x();
qreal y = accelReading.y();
qreal z = accelReading.z();
```

Here are some data explanations for common sensors:

Sensor reading	Values	Unit	Description
`QAccelerometerReading`	x, y, z	ms^2, meters per second squared	Linear acceleration along x, y, z axis
`QAltimeterReading`	altitude	Meters	Meters above average sea level
`QAmbientLightReading`	`lightLevel`	Dark, Twilight, Light, Bright, Sunny	General light level
`QAmbientTemperatureReading`	temperature	Celsius	Degrees Celsius
`QCompassReading`	azimuth	Degrees	Degrees from magnetic north
`QGyroscopeReading`	x, y, z	Degrees per second	Angular velocity around the axis
`QHumidityReading`	`absoluteHumidity`, `relativeHumidty`	gm^3, grams per cubic meter	Water vapor in air
`QIrProximityReading`	reflectance	Decimal fraction 0 - 1	How much infrared light was returned
`QLidReading`	`backLidClosed`, `frontLidClosed`	Bool	Laptop lid
`QLightReading`	lux	Lux	Light measured in lux
`QMagnetometerReading`	x, y, z	Magnetic flux density	Raw flux
`QOrientationReading`	orientation	TopUp, TopDown, LeftUp, RightUp, FaceUp, FaceDown	Enum device orientation
`QPressureReading`	pressure, temperature	Pascals, Celsius	Atmospheric pressure
`QProximityReading`	close	Bool	Close or far away
`QRotationReading`	x, y, z	Degrees	Rotation around axis in degrees

Some of these have sensor-specific readings, such as `QCompass` and `QMagnetometer`—both contain calibration levels.

Of course, C++ is not the only way to implement Qt's sensors; you can use them in QML as well. Let's find out how.

QML sensors

Of course, you can also use Qt Sensors from QML. In a lot of ways, it is easier to implement them this way, as the Qt Quick API has been optimized and simplified, so it takes less time to get the sensor up and running. Following our previous use of the light sensors, we will continue here. First off is the ever-present `import` statement: instead of calling a `start()` function to get it rolling, there is an `active` property. Instead of a `Lux` value, the property is `illuminance`. Not quite sure why there's a difference, but there you go:

```
import QtSensors 5.12
LightSensor {
    id: lightSensor
    active: true
    onReadingChanged {
        console.log("Lux "+ illuminance);
    }
}
```

It cannot get much simpler than that. `QtSensors` QML has no filter, so if you need to filter anything, you will have to use C++.

Custom QSensor and the backend engine

I want to briefly touch on how to create a custom sensor and engine backend. If you are on an embedded device, Qt Sensors may not have support for your sensor if it is a moisture or an air-quality sensor. You would need to implement your own `QSensor` and `QSensorBackend` engine.

There is a script in the directory, `src/sensors/make_sensor.pl`, that you can run which will generate a simple `QSensor` derived class, but additionally this script will generate Qt Quick classes that derive from `QmlSensor` . The `make_sensor.pl` script needs to be run from the `src/sensors` directory. For this exercise, we are going to create a sensor for monitoring salt concentrations in our saltwater swimming pool, so the name will be `QSaltSensor`.

You can then open these files in an editor, such as Qt Creator, and add what you need. Having a new `QSensor` type will also require a new backend that reads from the sensor.

Custom QSensor

There is a helper script named `QtSensors/src/sensors/make_sensor.pl`, which will create a basic template for a new `QSensor`, `QSensorReading`. It generates a simple `QSensor` derived class, but also classes for `QmlSensor` derived classes.

If you do not have it in your source directory, it can be found in the Git repository at `https://code.qt.io/cgit/qt/qtsensors.git/tree/src/sensors/make_sensor.pl`.

The `make_sensor.pl` script needs to be run from the `src/sensors` directory.

You will have to edit the resulting files and fill in a thing. For this example, I chose `QSaltSensor` as a class name. Execute the script with the class name as the first argument: `make_sensor.pl QSaltSensor`.

It creates the following files:

- `<sensorname>.h`
- `<sensorname>.cpp`
- `<sensorname>_p.h`
- `imports/sensors/<sensorname>.h`
- `imports/sensors/<sensorname>.cpp`

The output of using the `make_sensor.pl` command will appear like this:

```
cd src/sensors
$perl ./make_sensor.pl QSaltSensor
Creating ../imports/sensors/qmlsaltsensor.h
Creating ../imports/sensors/qmlsaltsensor.cpp
Creating qsaltsensor_p.h
Creating qsaltsensor.h
Creating qsaltsensor.cpp
You will need to add qsaltsensor to the src/sensors.pro file to the SENSORS
and the qmlsaltsensor files to src/imports/sensors/sensors.pro
```

Like the output says, you will need to add `qsaltsensor` to the `src/sensors.pro` file to the `SENSORS` variable that is used there. Add the `qmlsaltsensor` filepaths in the file, `src/imports/sensors/sensors.pro`.

Start by editing `qsaltsensor.cpp`, which is the class we will use as our `QSensorBackend`. The `perl` script we used to create the template has added comments where you should edit to customize. You will also need to add any properties.

Custom QSensorBackend

There are many reasons why you might want to implement your own sensor backend. One of these might be if you have a new type of sensor.

You would need to start implementing a backend engine for your new `QSensor` type. Begin by deriving from `QSensorBackend`:

```
#ifndef LINUXSALT_H
#define LINUXSALT_H

#include <qsensorbackend.h>
#include <qsaltsensor.h>

class LinuxSaltSensor : public QSensorBackend
{
```

The class `QSensorBackend`, has two pure virtual functions you need to implement: `start()` and `stop()`:

```
public:
    static char const * const id;
    LinuxSaltSensor(QSensor *sensor);
    ~LinuxSaltSensor();
    void start() override;
    void stop() override;
private:
    QSaltReading m_reading;
};
#endif // LINUXSALT_H
```

 The source code can be found on the Git repository under the `Chapter07-2` directory, in the `cp7` branch.

Implementing the backend functionality is up to you, based on if you have a salt sensor device you want to use. Of course you will have to compile and deploy your own Qt Sensors when you do so.

 For more information about custom `QSensors` and the backend, look at the Grue sensor example in Qt sensors. There is some rather amusing documentation on how to implement a custom sensor at `https://doc.qt.io/qt-5/qtsensors-grue-example.html`

Sometimes there are more than one sensor plugin on a system for any sensor. In this case we will need to tell the system which sensor to use. Let's look at how to configure QSensors.

Sensors.conf

If there is more than one backend for a particular sensor, you might need to specify which is the default.

You can add qsaltsensor to the Sensors.conf configuration file so the system can determine which sensor type is the default for a certain sensor. Of course, developers are free to choose whichever registered sensor on the system they want to use. The config file's format is SensorType = sensor.id, where SensorType is the base sensor class name, such as QLightSensor, and sensor.id is a String identifier for the specific sensor backend. The following code uses our saltsensor from the Linux backend and the list sensor from the sensorfw backend:

```
[Default]
QSaltSensor = linux.saltsensor
QLightSensor = sensorfw.lightsensor
```

QSensorGesture

QSensorGesture is an API for device gestures using sensors. As I mentioned in the introduction, they use ad-hoc gestures, which is to say there is no machine learning involved. Qt Sensors offers the following already-baked gestures:

- cover
- doubletap
- freefall
- hover
- pickup
- slam
- shake
- turnover
- twist
- whip

Instructions on how to perform specific gestures in Qt Sensors are detailed at http://doc. qt.io/qt-5/sensorgesture-plugins-topics.html.

It is worth noting that QSensorGesture uses signals specific to the recognizer. The slam gesture has the slam() signal, which gets emitted when it detects the slam gesture. It also has the standard detected("<gesture>") signal. The shake2 gesture has the shakeLeft, shakeRight, shakeUp and shakeDown signals, but also the corresponding detected signals.

The QSensorGesture class does not have the Q_OBJECT macro, and creates its signals at runtime directly on the meta object. As such, qobject_cast and subclassing QSensorGesture while using Q_OBJECT will not work.

QSensorGestureManager has the recognizerSignals function, which takes a gestureId so you can discover signals specific to the gesture, if you need to.

The source code can be found on the Git repository under the Chapter07-3 directory, in the cp7 branch.

To use QSensorGestures, create a QSensorGesture object, which takes a QStringList argument of a list of gesture IDs you want to use. You can specify directly which gestures you want using a QStringList like this:

```
QSensorGesture *gesture = new QSensorGesture(QStringList() <<
"QtSensors.slam", this);
connect(gesture, SIGNAL(detected(QString)), this,
SLOT(detectedGesture(QString)));
```

Alternatively you can also use QSensorGestureManager to get a list of all the registered gestures, calling gestureIds().

Because of the atypical implementation of QSensorGesture (because the signals get dynamically created at runtime), using the new style connect syntax, connect(gesture, &QSensorGesture::detected, this, &SomeClass::detectedGesture);, will result in a compiler error, as the new style syntax has compile-time checks.

Once you have these signals connected correctly, you call startDetection() for QSensorGesture:

```
gesture->startDetection();
```

QSensorGestureManager

You can get a list of all sensor gestures registered on the system by using
`QSensorGestureManager`:

```
QSensorGestureManager manager;

    for (const QString gestureId :  manager.gestureIds()) {
        qDebug() << gestureId;

      QStringList recognizerSignals =
manager.recognizerSignals(gestureId);

        for (const QString signalId : recognizerSignals ) {
            qDebug() << " Has signal " << signalId;
        }
  }
```

You can use the `gestureId` from the preceding code to create a new `QSensorGesture`
object and connect to the detected signal:

```
QSensorGesture *gesture = new QSensorGesture(QStringList() << gestureId,
this);
        connect(gesture,SIGNAL(detected(QString)),
this,SLOT(detectedGesture(QString)));
```

SensorGesture

Of course, sensor gestures can be used from QML. The API is slightly different in that there
is only one type, `SensorGesture`, so it is like using the generic `QSensor` class, except that,
instead of one gesture per object, `SensorGesture` can represent one or more gestures.

`SensorGesture` does not have its own import, and is lumped into `QtSensors`, so we need
to use that to indicate we are using the `QtSensors` module:

```
import QtSensors 5.12
```

You specify which gestures you want by writing to the `gestures` property, which takes a
list of strings of the `id` recognizer:

```
SensorGesture {
    id: sensorGesture
    gestures : [ "QtSensors.slam", "QtSensors.pickup" ]
}
```

Since there is only one generic `SensorGesture`, there are no gesture-specific signals. The gesture signal is `onDetected`, and a string of which gesture was detected is set in the `gesture` property. You will have to use some logic to filter for a certain gesture if you are using the component for more than one gesture:

```
onDetected: {
    if (gesture == "slam") {
        console.log("slam gesture detected!")
    }
}
```

To start the detection, write `true` to the `SensorGesture` enabled property:

```
sensor.gesture.enabled
```

You can grab your device and perform the slam gesture as outlined in the Qt documentation at `http://doc.qt.io/qt-5/sensorgesture-plugins-topics.html`. Depending on your device, it will detect a slam.

WebSockets – Bi-directional web communication

Now we are starting to get into the realm of network and the internet. WebSockets are a protocol that allows two-way data exchange between a web browser or client and a server without polling. You can stream data or send data at any time. Qt has support for WebSockets through the use of the `QWebSocket` API. Like normal TCP sockets, `QWebSockets` needs a server.

QWebSocketServer

`QWebSocketServer` can work in two modes: non-secure and SSL. We start by adding `websockets` to the `.pro` file so qmake sets up the proper library and header paths:

```
QT += websockets
```

Then include the `QWebSocketServer` header file:

```
#include <QtWebSockets/QWebSocketServer>
```

The source code can be found on the Git repository under the Chapter07-3 directory, in the cp7 branch.

To create a QWebSocketServer, it takes a server name as a string, a mode, and a parent object. The mode can be SecureMode or NonSecureMode.

SecureMode is is like HTTPS, uses SSL, and the protocol is wss. NonSecureMode is like HTTPS with the ws protocol:

```
const QWebSocketServer *socketServer = new
QWebSocketServer("MobileSocketServer",
QWebSocketServer::NonSecureMode,this);
connect(sockerServer, &QWebSocketServer::newConnection, this,
&SomeClass::newConnection);
connect(sockerServer, &QWebSocketServer::closed, this, &SomeClass::closed);
```

Like QSocket, there is a newConnection signal that gets emitted when a client attempts to connect to this server. If you are using SecureMode, you will want to connect to the sslErrors(const QList<QSslError> &errors) signal. Once the signals you want to use are connected, call listen to start the server, with a QHostAddress and a port number. QHostAddress::Any will listen to all network interfaces. You can specify one interface's address. The port number is optional and a port of 0 will be assigned a port automatically:

```
socketServer->listen(QHostAddress::Any, 7532);
```

Now we have a QWebSocketServer object that listens to incoming connections. We can handle this much like we did with the QSocketServer using nextPendingConnection in the corresponding slot we used to connect with the newConnection signal. That will give us a QWebSocket object that represents the connecting client.

QWebSocket

When a new connection comes in, QWebSocketServer emits the newConnection signal, which, here, calls the newConnection slot. We grab QWebSocket using the nextPendingConnection of the server object. With this, we connect to the QWebSocket signals:

```
QWebSocket *socket = socketServer->nextPendingConnection();
```

The first signal I like to connect is the error signal, as it can help debug. Like the QBluetooth class, the error function is overloaded, so special syntax is needed to use this signal.

The QWebSocket error signal is overloaded, so it needs unique handling to compile. QOverload is what you need to use.

```
connect(socket,
QOverload<QAbstractSocket::SocketError>::of(&QWebSocket::error),
        this, &SomeClass::socketError);
```

There are two types of messages that can be sent and received: text and binary. We have to deal with those differently, so there are signals for each. They get emitted by the server when the client sends a text or binary message:

```
connect(socket, &QWebSocket::textMessageReceived,
    this, &SomeClass::textMessageReceived);
connect(socket, &QWebSocket::binaryMessageReceived,
    this, &SomeClass::binaryReceived);
```

One difference between binary and text messages in WebSockets is that the text messages are terminated with the 0xFF character.

The textMessageReceived signal sends a QString, while the binaryMessageReceived sends a QByteArray:

```
SomeClass:binaryMessageReceived(const QByteArray &message) {
}
SomeClass:textMessageReceived(const QString &message) {
}
```

They also work on the frame level, but we are simply handling the entire message. If you have continuous streaming data of some kind, you might choose the textFrameReceived or binaryFrameReceived signals.

Client

A WebSocket client would simply use QWebSocket and connect to a server that supports WebSockets. One use case would be a web page (client) that shows sensor data sent though a QWebSocketServer.

QtQuick

Of course, the QWebSockets API provides QML components – WebSocket and WebSocketServer to be exact. As usual, it is quicker than using C++.

WebSocketServer

Add the following import line for your qml file to use WebSockets:

 The source code can be found on the Git repository under the Chapter07-4 directory, in the cp7 branch.

```
import QtWebSockets 1.0
```

To start listening with WebSocketServer, set the listen property to true. The url property, which takes a string, can be set to the address that clients will connect to:

```
WebSocketServer {
    id: socketServer
    url : "ws://127.0.0.1:33343"
    listen: true
}
```

When a client connects, the onClientConnected signal gets emitted, and its webSocket property represents the incoming WebSocket client. You also want to be able to do error checking so WebSocketServer has the onErrorStringChanged signal, with the errorString property. To do so, in the WebSocketServer component, implement it like this.

```
onClientConnected {
    ...
}

onErrorStringChanged {
    console.log(errorString)
}
```

Let's see how to handle the WebSocket for both server and client.

WebSocket

Both the client and server use WebSocket element. In the server, as I outlined in the *WebSocketServer* section, the client's `WebSocket` object can be obtained via the `onClientConnect` signal.

Check out how this works in the `WebSocketServer` component, as compared to the client:

```
WebSocketServer {
        id: socketServer
        host : "127.0.0.1"
        port: 33343
        listen: false
        onClientConnected {
            webSocket.onTextMessageReceived.connect(function(message)
    {
                console.log(message)
            });
        }
    }
```

The client requires the `url` property to be populated so it knows which server it will connect to:

```
WebSocket {
        id: webSocket
        url: "ws://localhost"

        onTextMessageReceived {
            console.log(message)

        }
    }
```

The incoming message appears in the `onTextMessageReceived` signal with the `message` property.

To send a message to the server or client, `WebSocket` has the `sendMessage` function. If this is the server, the `webSocket` would be used to send a message of text like this.

```
webSocket.sendTextMessage("socket connected ok!")
```

WebSockets for Qt Quick does not handle binary messages in the true sense of the word. It does happen to have an `onBinaryMessageReceived` signal, but the `message` object that gets received is a `String`. I would suggest that if your binary message will get messed up by being converted to UTF16-encoded `QString`, you might consider using the C++ API.

QMqtt – Brokers of machine talk

MQTT is a publish-and-subscribe messaging transport. There was a similar framework in the Qt Mobility stack called Publish and Subscribe, which is now part of the officially unsupported `QSystems` API framework, which also includes `QSystemInfo` and `QSystemFramework`.

`QMqtt` is a framework for writing MQTT clients. You will need to install and run an MQTT broker, such as Mosquitto or HiveMQ, or use an internet-based service. For my development and testing purposes, I chose HiveMQ. You can download it from `https://www.hivemq.com/`.

They also have a public broker at `http://www.mqtt-dashboard.com/index.html`.

MQTT has a broker, or server that one or more clients connect to. The clients can then publish and/or subscribe to different topics.

You can use `QWebSockets` to access a broker, and there is an example in Qt, which uses the `WebSocketIODevice` class in the `examples/mqtt/websocketsubscription` directory.

QMqttClient

To start developing a `QMqttClient`, you will have to build it yourself, as it does not get distributed with Qt itself, unless you get the commercial Qt for Automation.

You download the open source licensed version from `git://code.qt.io/qt/qtmqtt.git`.

Luckily it is a straightforward and easy build. Once you run `qmake;` `make && make install;`, you are ready to use it.

In your `pro` file, we need to add the `mqtt` module.

```
QT += mqtt
```

The header file is named `QtMqtt/QMqttClient`, so let's include that:

```
#include <QtMqtt/QMqttClient>
```

The source code can be found on the Git repository under the `Chapter07-5` directory, in the `cp7` branch.

The main class we use to access the broker is named `QMqttClient`. It can be thought of as the manager. It has a simple construction. You need to specify the host and port with the `setHostname` and `setPort` functions. We will use the `hivemq` public broker and port `1883`:

```
mqttClient = new QMqttClient(this);
mqttClient->setHostname(broker.hivemq.com);
mqttClient->setPort(1883);
```

It is a good idea connect to any error signals to help debugging when things go wrong; let's do that first:

```
connect(mqttClient, &QMqttClient::errorChanged, this,
&SomeClass::errorChanged);
connect(mqttClient, &QMqttClient::stateChanged, this,
&SomeClass::stateChanged);
connect(mqttClient, &QMqttClient::messageReceived, this,
&SomeClass::messageReceived);
```

To connect to the `mqtt` broker, call `connectToHost();`:

```
mqttClient->connectToHost();
```

Since we connected to the `stateChanged` signal, we can wait until we are connected to the broker to subscribe to any topics:

```
void SomeClass::stateChanged(QMqttClient::ClientState state)
{
    switch(state) {
     case QMqttClient::Connecting:
         qDebug() << "Connecting...";
         break;
     case QMqttClient::Connected:
         qDebug() << "Connected.";
         subscribe();
         break;
     case QMqttClient::Disconnected:
         qDebug() << "Disconnected."
         break;
    }
}
```

The QMqttClient::subscribe function takes a topic in the form of QMqttTopicFilter. Here, I assign it the "Qt" string.

It returns a QMqttSubscription pointer, which we can use to connect to the stateChanged signal. We will then simply subscribe to the topic we just published.

Our subscribe function looks like this:

```
void MainWindow::subscribe()
{
    QMqttTopicFilter topicName("Qt");
    subscription = mqttClient->subscribe(topicName, 0);
    connect(subscription, &QMqttSubscription::stateChanged, this,
            &SomeClass::subscriptionStateChanged);
    publish();
}
```

We simply call our function that will then publish something to that topic.

QMqttClient::publish takes a topic name in the form of a QMqttTopicName, and the message is just a standard QByteArray.

The publish function looks like this:

```
void MainWindow::publish()
{
    QMqttTopicName topicName("Qt");
    QByteArray topicMessage("Everywhere!");
    mqttClient->publish(topicName, topicMessage);
}
```

You should then see the message we published in the messageReceived slot:

Putting it all together

I have a Raspberry Pi and a Sense HAT board that I can use to collect sensor data. Luckily, I previously wrote a Qt Sensors plugin for the Sense HAT. It happens to be in a standalone Git repository and not in any Qt Sensors version, unlike the TI SensorTag backend plugin.

If you don't want to write your own Sense HAT sensor plugin you can get my standalone Sense HAT plugin from `https://github.com/lpotter/qsensors-sensehatplugin.git`.

The version of Qt Sensors on the Raspbian distribution is 5.7 and does not have the pressure and humidity sensors that the Sense HAT has. They were added in later Qt Sensors versions.

Cross-compiling on a desktop is so much faster than compiling on the device—days on the **Raspberry Pi (rpi)** as opposed to a few minutes on a good development machine. I had some trouble getting the cross-compiling `toolchain` to work, so I had to opt for the on-board native compile, which of course takes a very long time on a Raspberry Pi. The easiest way is to get Qt's commercial `Boot2Qt` and `Automation` packages, as they package it up nicely, and provide binaries and support.

Since this book uses Qt 5.12, we need to get the explicit version of the following Qt module repositories, by using the following Git commands:

- Qt Base: `git clone http://code.qt.io/qt/qtbase.git -b 5.12`
- Qt WebSockets: `git clone http://code.qt.io/qt/qtwebsockets.git -b 5.12`
- Qt MQTT: `git clone http://code.qt.io/qt/qtmqtt -b 5.12`
- Qt Sensors: `git clone http://code.qt.io/qt/qtsensors -b 5.12`

We are going to create an app for Raspberry Pi that grabs the Sense HAT's temperature and pressure data and distributes them via `QMqtt` and `QWebSockets` to a broker running on HiveMQ.

 The source code can be found on the Git repository under the `Chapter07-6` directory, in the `cp7` branch.

Start by implementing a `SensorServer` class, which is typically a `QObject` derived class.

```
SensorServer::SensorServer(QObject *parent)
  : QObject(parent),
    humiditySensor(new QHumiditySensor(this)),
    temperatureSensor(new QAmbientTemperatureSensor(this))
{
    initSensors();
    initWebsocket();
}
```

We then implement the `QWebSockeIODevice` that we declared as `mDevice` and connect to its `socketConnected` signal.

```
void SensorServer::initWebsocket()
{
    mDevice.setUrl("broker.hivemq.com:8000");
    mDevice.setProtocol("mqttv3.1");

    connect(&mDevice, &WebSocketIODevice::socketConnected, this,
&SensorServer::websocketConnected);
}
```

Next we call the `connectToBackend()` function of the sensors we want to use.

```
void SensorServer::initSensors()
{
    if (!humiditySensor->connectToBackend()) {
        qWarning() << "Could not connect to humidity backend";
    } else {
        humiditySensor->setProperty("alwaysOn",true);
        connect(humiditySensor,SIGNAL(readingChanged()),
                this, SLOT(humidityReadingChanged()));
    }
    if (!temperatureSensor->connectToBackend()) {
        qWarning() << "Could not connect to humidity backend";
    } else {
        temperatureSensor->setProperty("alwaysOn",true);
        connect(temperatureSensor,SIGNAL(readingChanged()),
                this, SLOT(temperatureReadingChanged()));
    }
}
```

The call to `initSensors()` connects to the sensor's backend and sets up `readingChanged` signal connections.

To use `QWebSockets` for the `QMqtt` client, we need to create a `QIODevice` that uses `QWebSockets`. Luckily, there is one already written in the `QMqtt` `examples/mqtt/websocketssubscription` directory, named `websocketsiodevice`, so we will import that into the project:

```
SOURCES += websocketiodevice.cpp
HEADERS += websocketiodevice.h
```

In our header file, we include `websocketdevice.h`.

```
#include "websocketiodevice.h"
```

In the class declaration, we instantiate the `WebSocketIODevice` as a class member.

```
WebSocketIODevice mDevice;
```

To actually use `WebSocketIODevice`, we need to set it as the `QMqttClient` transport.

We first set up our `WebSocketIODevice` and connect to its `socketConnected` signal to set up `QMqtt`.

The `mqtt` broker at `hivemq` uses a different port number, so we set it in the URL:

```
void SensorServer::initWebsocket()
{
    mDevice.setUrl(QUrl("broker.hivemq.com:8000"));
    connect(&mDevice, &WebSocketIODevice::socketConnected, this,
&SensorServer::websocketConnected);
    mDevice.open(QIODevice::ReadWrite);
}
```

Now we set up `QMqtt` and set its transport to use `WebSocketIODevice`. We are using a transport with its own connection, so we do not set the URL for the `QMqtt` object, but rely on the `websocket` for connection. We then set up `mqttClient` as usual:

```
void SensorServer::websocketConnected()
{
    mqttClient = new QMqttClient(this);
    mqttClient->setProtocolVersion(QMqttClient::MQTT_3_1);
    mqttClient->setTransport(&mDevice, QMqttClient::IODevice);
    connect(mqttClient, &QMqttClient::errorChanged,
            this, &SensorServer::errorChanged);
    connect(mqttClient, &QMqttClient::stateChanged,
            this, &SensorServer::stateChanged);
```

```
connect(mqttClient, &QMqttClient::messageReceived,
        this, &SensorServer::messageReceived);

mqttClient->connectToHost();
}
```

We monitor the changing state and act when it becomes `Connected`. We will start the `humidity` and `temperature` sensor, and then call subscribe so we can monitor when the `mqtt` broker is publishing:

```
void SensorServer::stateChanged(QMqttClient::ClientState state)
{
    switch(state) {
    case QMqttClient::Connecting:
        qDebug() << "Connecting...";
        break;
    case QMqttClient::Connected:
        qDebug() << "Connected.";
        humiditySensor->start();
        temperatureSensor->start();
        subscribe();
        break;
    case QMqttClient::Disconnected:
        qDebug() << "Disconnected.";
        break;
    }
}
```

In our sensor's `readingChanged` slots, we will publish the data to the `mqtt` broker:

```
void SensorServer::humidityReadingChanged()
{
    qDebug() << Q_FUNC_INFO << __LINE__;
    QHumidityReading *humidityReading = humiditySensor->reading();
    QByteArray data;
    data.setNum(humidityReading->relativeHumidity());
    QMqttTopicName topicName("Humidity");
    QByteArray topicMessage(data);
    mqttClient->publish(topicName, topicMessage);
}

void SensorServer::temperatureReadingChanged()
{
    qDebug() << Q_FUNC_INFO << __LINE__;
    QAmbientTemperatureReading *tempReading = temperatureSensor
>reading();
    QByteArray data;
    data.setNum(tempReading->temperature());
```

```
    QMqttTopicName topicName("Temperature");
    QByteArray topicMessage(data);
    mqttClient->publish(topicName, topicMessage);
}
```

Finally, let's see any subscribed messages:

```
void SensorServer::messageReceived(const QByteArray &message, const
QMqttTopicName &topic)
{
    qDebug() << Q_FUNC_INFO  << topic << message;
}
```

Summary

In this chapter, we looked at the different ways of using `QSensors` to read a device's sensor data. There are many supported platforms for Qt Sensors: Android, iOS, WinRT , SensorTag, Sensorfw, Linux generic, and Linux iio-sensor-proxy. Sensorfw also has support for Linux's IIO sensors.

I described how to implement custom `QSensor` and `QSensorBackend` to add support for sensors not currently supported in Qt Sensors.

We went through the steps involved in using `QtMqtt` to talk to an `mqtt` broker, and we looked at how to use `QWebsockets` to communicate to a web server web page.

Then I threw them all together to grab sensor data from a Sense HAT, and publish them to `mqtt` broker by way of WebSockets.

In the next chapter, we will discuss using GPS data comprising of location and position and mapping.

8
Where Am I? Location and Positioning

Devices with GPS chips are everywhere. You can even track your cat or chicken! In this chapter, you will learn how to use Qt for location and positioning services.

Qt Positioning entails geographic coordinates from various sources, including satellites, Wi-Fi, and log files. Qt Location is all about local places, for example services, such as restaurants or public parks, and also routing information.

In this chapter, we will cover the following topics:

- Positioning with satellites
- Mapping the positions
- Places of interest

Positioning with satellites

A phone usually has a built-in GPS modem but also other sources of positioning information, so I will use Android for this example. The main Qt classes we will look at are as follows:

Here are the Qt Positioning APIs:

- QGeoSatelliteInfo
- QGeoLocation
- QGeoPositionInfoSource
- QGeoCoordinate

And here are the Qt Location APIs:

- QPlaceSearchResult
- QPlaceContent
- QGeoRoute

First, we need to edit the .pro file and add QT += positioning.

QGeoSatelliteInfoSource

You can show the user satellite information by using QGeoSatelliteInfoSource, which has a static method to get QGeoSatelliteInfoSource.

 The source code can be found on the Git repository under the Chapter08-1 directory, in the cp8 branch.

We will start by calling QGeoSatelliteInfoSource::createDefaultSource.

```
QGeoSatelliteInfoSource *source =
QGeoSatelliteInfoSource::createDefaultSource(this);
```

 On some systems, such as iOS, satellite information is not exposed to the public API, so QGeoSatelliteInfoSource will not work on that platform.

This constructs a QGeoSatelliteInfoSource object for the highest-priority plugin on the system, which is about the same as doing the following:

```
QStringList geoSources = QGeoSatelliteInfoSource::availableSources();
QGeoSatelliteInfoSource *source =
QGeoSatelliteInfoSource::createSource(geoSources.at(0),this);
```

There are two signals of particular interest: satellitesInUseUpdated and satellitesInViewUpdated. In addition, there is the overloaded error signal, so we need to use the special QOverload syntax:

```
connect(source, QOverload<QGeoSatelliteInfoSource::Error>::
    of(&QGeoSatelliteInfoSource::error),
    this, &SomeClass::error);
```

The `satellitesInUseUpdated` signal is emitted when the number of satellites that the system is using changes. The `satellitesInViewUpdated` signal gets emitted when the number of satellites the system can see changes. We will receive a list of `QGeoSatelliteInfo` objects.

QGeoSatelliteInfo

Let's connect the `satellitesInViewUpdated` signal so we can detect when satellites are found:

```
connect(source, SIGNAL(satellitesInViewUpdated(QList<QGeoSatelliteInfo>)),
    this, SLOT(satellitesInViewUpdated(QList<QGeoSatelliteInfo>)));
```

We can receive information for individual satellites this way. Information such as a satellite identifier, signal strength, elevation, and azimuth is included:

```
void SomeClass::satellitesInViewUpdated(const QList<QGeoSatelliteInfo>
&infos)
{
    if (infos.count() > 0)
        qWarning() << "Number of satellites in view:" << infos.count();

    foreach (const QGeoSatelliteInfo &info, infos) {
        qWarning() << "      "
            << "satelliteIdentifier" << info.satelliteIdentifier()
            << "signalStrength" << info.signalStrength()
            << (info.hasAttribute(QGeoSatelliteInfo::Elevation) ?
"Elevation "
+ QString::number(info.attribute(QGeoSatelliteInfo::Elevation)) : "")
            << (info.hasAttribute(QGeoSatelliteInfo::Elevation) ?  "Azimuth
" +
QString::number(info.attribute(QGeoSatelliteInfo::Azimuth)) : "");
    }
}
```

There's a lot to see here on a small screen. Every update is a new line, and you can see as it locates and adds different satellites when they come into view:

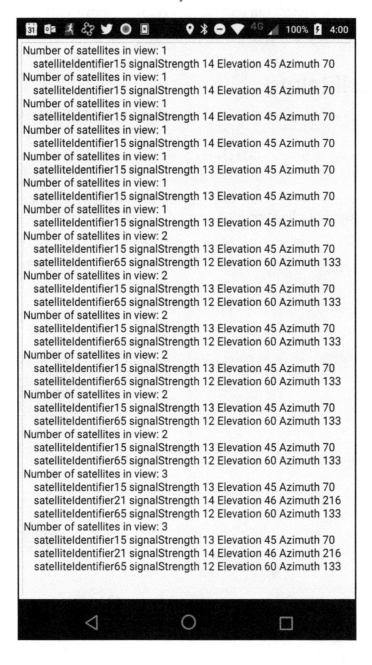

The next step is to use those satellites to triangulate our position on the globe. We start by using QGeoPositionInfoSource.

QGeoPositionInfoSource

We can get the latitude and longitude position of the device by using QGeoPositionInfoSource, which encapsulates positional data. Like QGeoSatelliteInfoSource, it has two static methods to create the source object:

 The source code can be found on the Git repository under the Chapter08-2 directory, in the cp8 branch.

```
QGeoPositionInfoSource *geoSource =
QGeoPositionInfoSource::createDefaultSource(this);
```

The QGeoPositionInfoSource signal we are interested in is positionUpdated(const QGeoPositionInfo &update):

```
connect(geoSource, &QGeoPositionInfoSource::positionUpdated,
    this, &MainWindow::positionUpdated);
```

To start receiving updates, call startUpdates();:

```
geoSource->startUpdates();
```

The positionUpdated signal receives a QGeoPositionInfo.

QGeoPositionInfo

QGeoPositionInfo contains a QGeoCoordinate that contains our latitude and longitude coordinates, as well as a timestamp for the location data.

It can also contain the following optional attributes:

- Direction
- GroundSpeed
- VerticalSpeed
- MagneticVariation

- `HorizontalAccuracy`
- `VerticalAccuracy`

The attributes can be checked with `hasAttribute(QGeoPositionInfo::Attribute)` and retrieved with the `attribute(QGeoPositionInfo::Attribute)` function:

```
if (positionInfo.hasAttribute(QGeoPositionInfo::MagneticVariation)
    qreal magneticVariation =
positionInfo.attribute(QGeoPositionInfo::MagneticVariation);
```

To get latitude and longitude information, call the `coordinate()` function in the `QGeoPositionInfo` class, which returns a `QGeoCoordinate`.

QGeoCoordinate

`QGeoCoordinate` contains the latitude and longitude coordinates, and can be found calling the respective `latitude()` and `longitude()` functions. It can be made up of different types of data, and can be discovered by calling the `type()` function, which returns an `enum` of `QGeoCoordinate::CoordinateType`, which can be one of the following values:

- `InvalidCoordinate`: Invalid coordinate
- `Coordinate2D`: Contains latitude and longitude coordinates
- `Coordinate3D`: Contains latitude, longitude, and altitude coordinates

We can get the `QGeoCoordinate` from the `QGeoPositionInfo` object's `coordinate()` function which, in turn, has `latitude` and `longitude` values:

```
QGeoCoordinate coords = positionInfo.coordinate();
QString("Latitude %1\n").arg(coords.latitude());
QString("Longitude %1\n").arg(coords.longitude());

if (coords.type() == QGeoCoordinate::Coordinate3D)
    QString("Altitude %1\n").arg(coords.altitude())
```

Let's take a look at how we do this using Qt Quick and QML.

Qt Quick

There are corresponding QML elements available for positioning.

The `import` statement would be `import QtPositioning 5.12`.

Let's do the same simple thing with QML and show our latitude and longitude values.

Here are the Qt Quick item equivalents of the previously-mentioned classes:

- `PositionSource: QGeoPositionInfoSource`

- `Position: QGeoPositionInfo`

- `Coordinate: QGeoCoordinate`

Qt Quick is often much simpler, and quick to implement these things.

 The source code can be found on the Git repository under the `Chapter08-3` directory, in the `cp8` branch.

We implement `PositionSource` with an `updateInterval` of 1,000, which means the devices position will update every 1,000 milliseconds. We set it to `active` to start the updates:

```
PositionSource {
    id: positionSource
    updateInterval: 1000
    active: true
```

This component has a signal named `onPositionChanged`, which gets called when the position changes. We receive the changed coordinates and can then use them:

```
onPositionChanged: {
    var coord = positionSource.position.coordinate;
    console.log("Coordinate:", coord.longitude, coord.latitude);
    latitudeLabel.text = "Latitude: " + coord.latitude;
    longitudeLabel.text = "Longitude: " + coord.longitude;
    if (positionSource.position.altitudeValid)
        altitudeLabel.text = "Altitude: " + coord.altitude;
    }
}
```

Now that we know where we are, we can use those location details to get certain details around the coordinates, like the map and location's place details.

Mapping the positions

We now need a map of some sort to show our location findings.

> The Map component for QML is the only way Qt provides for mapping, so you will have to use Qt Quick.

The Map component can be backed by various backend plugins. In fact, you need to specify which plugin you are using. Map has built-in support for the following plugins:

Provider	Key	Notes	Url
Esri	esri	Subscription required	www.esri.com
HERE	here	Access token required	developer.here.com/terms-and-conditions
Mapbox	mapbox	Access token required	www.mapbox.com/tos
Mapbox GL	mapboxgl	Access token required	www.mapbox.com/tos
Open Street Map (OSM)	osm	Free access	openstreetmap.org/

I will be using the OSM and HERE providers.

The HERE plugin requires an account at `developer.here.com`. It's easy to sign up and there is a free version. You need the app ID and app code to access their maps and API.

Map

To start using the Map component, in your chosen `.qml` file, add both `QtLocation` and `QtPositioning` in the `import` lines:

```
import QtLocation 5.12
import QtPositioning 5.12
```

The source code can be found on the Git repository under the `Chapter08-4` directory, in the `cp8` branch.

The `Map` component needs a `plugin` object, whose `name` property is one of the keys from the preceding table. You can set where the map is centered by setting the `center` property to a coordinate.

I am using the OSM backend and it is centered on the Gold Coast, Australia:

```
Map {
    anchors.fill: parent
    plugin: Plugin {
        name: "osm"
    }
    center: QtPositioning.coordinate(-28.0, 153.4)
    zoomLevel: 10
}
```

The Map centers on the coordinates we indicate with the `center` property, which is used to position the map to the user's current location.

We defined the Map's `plugin` property to be the `"osm"` plugin, which is an identifier for the Open Street Map plugin.

It is that easy to show a map.

MapCircle

You can highlight an area by placing a `MapCircle` in the `Map`. Again, centered on the Gold Coast.

A `MapCircle` has a `center` property that we can define by using a `latitude` and `longitude` location in a signed decimal value.

The `radius` property here is in the unit of meters according to the map. So in our example, the `MapCircle` will have a radius of 5,000 meters.

```
MapCircle {
    center {
        latitude: -28.0
        longitude: 153.4
    }
    radius: 5000.0
    border.color: 'red'
    border.width: 3
    opacity: 0.5
}
```

Each map backend has its own parameters, which can be set using the `PluginParameter` component in the `Map` component.

PluginParameter

By default, the OSM backend downloads lower-resolution tiles. If you want high-resolution maps, you can specify the `'osm.mapping.highdpi_tiles'` parameter:

```
PluginParameter {
    name: 'osm.mapping.highdpi_tiles'
    value: true
}
```

Each `PluginParameter` element holds just one `name`/`value` parameter pair. If you need to set several parameters, you will need a `PluginParameter` element for each:

```
PluginParameter { name: "osm.useragent"; value: "Mobile and Embedded
Development with Qt5"; }
```

Other `PluginParameters` you could consider are tokens and app IDs for various map providers, such as HERE maps.

Here's how our map looks, running on Android:

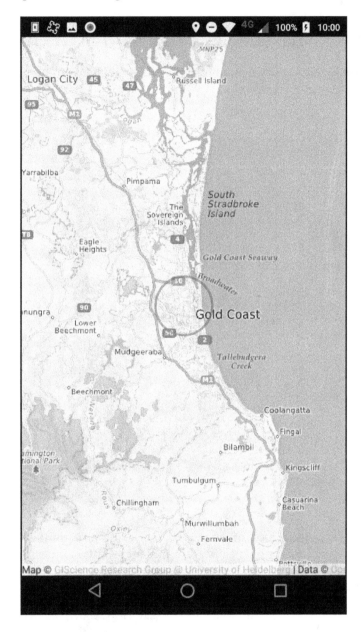

There are other Qt Quick items that we can use with addresses on the map. Let's look at routing.

RouteModel

To show a route on a map, you will need to use RouteModel, which is a property of the Map item, RouteQuery to add waypoints, and a MapItemView to display it.

RouteModel needs a plugin, so we just reuse the plugin for the Map item. It also needs a RouteQuery for its query property:

```
RouteQuery {
    id: routeQuery
}
RouteModel {
    id: routeModel
    plugin : map.plugin
    query: routeQuery
}
```

MapItemView is used to display model data on the map. It also needs a delegate of MapRoute. In our case, this is a line that describes the route:

```
MapItemView {
    id: mapView
    model: routeModel
    delegate: routeDelegate
}
Component {
    id: routeDelegate
    MapRoute {
        id: route
        route: routeData
        line.color: "#46a2da"
        line.width: 5
        smooth: true
        opacity: 0.8
    }
}
```

Now what we need is a starting point, an ending point, and any point in between. In this example, I keep it simple and only specify start and end points. You can specify a GPS coordinate by using QtPositioning.coordinate, which takes a latitude and longitude value as arguments:

```
property variant startCoordinate: QtPositioning.coordinate(-28.0, 153.4)
property variant endCoordinate: QtPositioning.coordinate(-27.579744,
153.100175)
```

The start-point coordinate is some random area on the Gold Coast, Australia; the endpoint is where the last south-of-the-equator Trolltech office was. The `RouteQuery` `travelModes` property determines how the route is figured, whether traveling by car, foot, or public transport. It can be one of the following values:

- `CarTravel`
- `PedestrianTravel`
- `BicycleTravel`
- `PublicTransit`
- `TruckTravel`

The `RouteQuery` **property**, `routeOptimzations`, limits the query to the following different values:

- `ShortestRoute`
- `FastestRoute`
- `MostEconomicRoute`
- `MostScenicRoute`

In this example, I made `routeQuery` fire off in the `Component.onCompleted` signal. Usually, something like this would be triggered after the user has configured the query:

```
Component.onCompleted: {
    routeQuery.clearWaypoints();
    routeQuery.addWaypoint(startCoordinate)
    routeQuery.addWaypoint(endCoordinate)
    routeQuery.travelModes = RouteQuery.CarTravel
    routeQuery.routeOptimizations = RouteQuery.FastestRoute
    routeModel.update();
}
```

Here is how the route looks. This route indicated by the blue line starting in the big red circle:

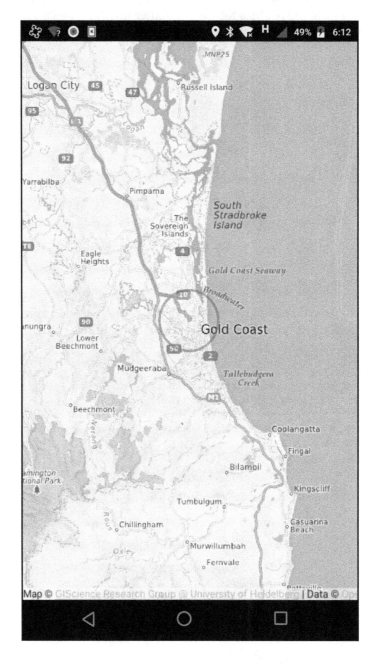

You can add more `Waypoints` to establish different routes or get turn-by-turn directions by setting `routeModel` to a `ListView` or similar.

Not only can Qt Location show maps and routes, but there is also support for displaying places of interest, such as restaurants, gas stations, and national parks, in the `Places` API.

Places of interest

At this point, I am going to switch to the HERE maps plugin. I tried to get the OpenStreetMaps places to work, but it could not find anything around.

In the next step of the construction of our map, we use `PlaceSearchModel` to search for places. As with the `RouteModel` before, `MapItemView` can display this model on our map.

Just like `RouteModel`, `PlaceSearchModel` needs some way of displaying the data; we could choose a `ListView`, which is useful for some purposes, but let's choose `MapItemView` for the visual effect.

We need to state which plugin we are using with `searchArea` and `searchTerm`:

```
PlaceSearchModel {
    id: searchModel
    plugin: mapPlugin
    searchTerm: "coffee"
    searchArea: QtPositioning.circle(startCoordinate)
    Component.onCompleted: update()
}
```

Our `MapItemView` and `delegate` code look like this. The `searchView` delegate will show up as an icon with its title text, from the resulting place :

```
MapItemView {
    id: searchView
    model: searchModel
    delegate: MapQuickItem {
        coordinate: place.location.coordinate
        anchorPoint.x: image.width * 0.5
        anchorPoint.y: image.height
        sourceItem: Column {
            Image { id: image; source: "map-pin.png" }
            Text { text: title; font.bold: true; color: "red"}
        }
    }
}
```

As you can see here, the place points are a bit difficult to read and are superimposed on top of the other ones that are around. This indicates that there are places too near each other for the zoom level and the map is having difficulties in placing the names. You can work around this issue by using a different zoom level or by using some collision detection and layout algorithms that I won't go into there.

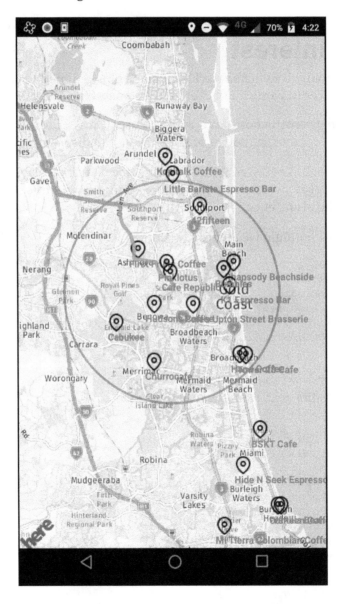

The `map-pin.png` icon is from `https://feathericons.com/` and is released under the open source MIT license.

Summary

In this chapter, we covered many aspects of mapping using Qt Location and Qt Positioning. We can get satellite information with `QGeoSatelliteInfo`, and locate the exact current position coordinates with `QGeoPositionInfo`. We learned how to use Qt Quick `Map` and different map providers to show the current location. We covered how to provide a route with `RouteModel`, search for places nearby using `PlaceSearchModel`, and show them using `MapItemView`.

In the next chapter, we will discuss audio and video with Qt Multimedia.

Section 3: Other APIs Qt SQL, Qt Multimedia, and Qt Purchasing

We will begin this section by discussing some useful APIs for mobile devices, such as audio and video in games. We'll then move on to the challenges posed by databases in the devices, and learn how to utilize the database remotely over a network. We'll also discuss how to enable in-app purchasing with Qt Purchasing.

This section comprises of the following chapters:

- Chapter 9, *Sounds and Visions - Qt Multimedia*
- Chapter 10, *Remote Databases with Qt SQL*
- Chapter 11, *Enabling In-App Purchases with Qt Purchasing*

Sounds and Visions - Qt Multimedia

9

Applications that need to play sounds or show videos are usually games, while others are full-blown multimedia apps. Qt Multimedia can handle both.

Qt Multimedia can be used with both Qt Widgets and Qt Quick, or even without a GUI interface. It has both C++ and QML APIs, but the QML API has a few special treats and tricks. A little-known fact is that Qt can also play 3D positional audio in Qt Quick. You can control the gain and pitch with three dimensions.

We will cover the following topics in this chapter:

- Sonic vibrations – audio
- Image sensors – camera
- Visual media – playing video
- Tuning it in – FM radio tuner

Sonic vibrations – audio

I have a relationship with audio that goes way back—before computers were household things, when Mylar tape and magnets ruled the sonic realms. Things have progressed since then. Now, mobile phones fit into our pockets and light bulbs can play music.

3D audio in Qt is supported through the OpenAL API. If you are using Linux, the default Qt binaries from Qt Company do not ship with the needed Qt Audio Engine API. You will have to install the OpenAL development package and then compile Qt Multimedia for yourself. OpenAL is not supported on Android, so no joy there. Luckily, it is supported by default on Apple Mac and iOS. So, that is where I am going to develop this next section. Let's grab the nearest MacBook and head over there.

3D audio is audio in three dimensions, just like 3D graphics—not just left and right, but also up, down, front, and back placement of audio. The term *positional audio* might explain this better.

With Qt, 3D audio is only supported using Qt Quick.

 The source code for this chapter can be found in the Git repository under the `Chapter09-3dAudio` directory, in the `cp9` branch.

To use Qt Multimedia, you need to edit the project `.pro` file and add the following line:

```
QT += multimedia
```

Edit the `qml` file that you want to use the 3D audio in, and add the `import` line:

```
import QtAudioEngine 1.0
```

3D space is made up of three axes named x, y, and z, which correspond to horizontal/vertical and up/down in 3-dimensional space.

`AudioEngine` and other associated classes use the `Qt.vector3d` value type. It is essential to understand this element to use 3D audio.

Qt.vector3d

`Qt.vector3d` is an array of values that represents the x, y, and z axes—x being horizontal, y being vertical, and z being up or down. Each value is a single-precision `qreal`.

It can be used like `Qt.vector3d(15, -5, 0)` or `"15, -5, 0"` as a `String`.

The positioning of the audio is controlled through the use of the `vector3d` property value.

`Qt.vector3d` is used to position the audio in 3 dimensional space.

The main component for using 3D audio in QML is called `AudioEngine`. The other components we will use can be children of this component.

AudioEngine

`AudioEngine` is the central container for the other 3D audio items that you will use.

We can set up the component easily enough:

```
AudioEngine {
    id: audioEngine
    dopplerFactor: 1
    speedOfSound: 343.33
}
```

The `dopplerFactor` property creates a Doppler shift effect. The `speedOfSound` value reflects the speed of sound in which the Doppler effect is calculated.

You assign a `listener` through the `listener` property. We will get to that later in the *AudioListener* section.

We have an audio sample we want to load and use, so we declare at least one `AudioSample`.

AudioSample

`AudioSample` can be defined as a child of an `AudioEngine` component:

```
AudioEngine {
    id: audioEngine
    dopplerFactor: 1
    speedOfSound: 343.33
    AudioSample {
        name:"plink"
        source: "thunder.wav"
        preloaded: true
    }
}
```

It can also be added using the `AudioEngine.addAudioSample()` method:

```
AudioEngine {
    id: audioEngine
    dopplerFactor: 1
    speedOfSound: 343.33
    addAudioSample(plinkSound)
}
AudioSample {
```

```
        id: plinkSound
        name:"plink"
        source: "thunder.wav"
        preloaded: true
    }
```

The `source` property holds the sample's filename and a name to refer to it with.

Now, we are ready to play the sound using the `Sound` component.

Sound

The `Sound` element is a container for one or more samples that will play with different parameters and variances. In other words, you can define a `PlayVariation` item, which defines how a `Sound` plays an `AudioSample`, with maximum or minimum values in pitch and gain. You can also declare the sample to be `looping`, which means it plays over and over:

```
Sound {
    name: "thunderengine"
    attenuationModel: "thunderModel"
    PlayVariation {
        looping: true
        sample: "plink"
        maxGain: 0.5
        minGain: 0.3
    }
}
```

The `attenuationModel` property controls the way the sound volume level falls off, or fades over time. It can be one of these values:

- Linear is a straight falloff
- Inverse is a more natural, non-linear curve

You can control this using the `start`, `end`, and `rolloff` properties.

AudioListener

The `AudioListener` component represents the `listener` and its position in the 3D realm. There is only one `listener`. It can either be constructed as the `listener` property of the `AudioEngine` component, or as a definable element:

```
AudioListener {
    engine: audioEngine
    position: Qt.vector3d(0, 0, 0)
}
```

A `SoundInstance` is the component that a `Sound` uses to play the sample.

SoundInstance

`SoundInstance` has a few properties that you can use to adjust the sound:

- direction
- gain
- pitch
- position

These properties take a `vector3d` value.

The `sound` property of the `SoundInstance` element takes a string that represents the name of a `Sound` component:

```
SoundInstance {
    id: plinkSound
    engine: audioEngine
    sound: "thunderengine"
    position: Qt.vector3d(leftRightValue, forwardBacktValue,
upDownValue)
    Component.onCompleted: plinkSound.play()
}
```

Here, I start playing the sound when the component is completed.

Now, we just need some mechanism to move the sound position around. We can use the `Accelerometer` values if we have an accelerometer on the device. I'm just going to use the mouse. Remember that on a touchscreen, a `MouseArea` also includes touch input.

We must enable `hover` in order to track the mouse without clicking:

```
MouseArea {
    anchors.fill: parent
    hoverEnabled: true
    propagateComposedEvents: true
    onPositionChanged: {
        leftRightValue = -((window.width / 2) - mouse.x)
        forwardBacktValue = (window.height / 2) - mouse.y
    }
```

To propagate the mouse clicks to buttons or other items when using `MouseArea`, put the `MouseArea` at the top of the file, as Qt Quick will set the z order in the order of the components from the top of the file, down to the bottom. You could also set the z property of the buttons and set the z property of the `MouseArea` to the lowest value.

I previously declared three values in my `Window` component to use in the positioning of the audio:

```
property real leftRightValue: 0;
property real forwardBacktValue: 0;
property real upDownValue: 0;
```

Now, when you move the mouse around, the audio will appear to move around.

But there is no mouse on a phone. There is a touch point, but no scrolling. I could use an `Accelerometer` as it has the z axis, or use `PinchArea` to control the up and down position.

Let's look at a few other ways to deal with audio.

Audio

The `Audio` element is probably the easiest way to play audio. It only takes a few lines. It would be good for playing sound effects.

The source code can be found in this book's Git repository under the `Chapter09-1` directory, in the `cp9` branch.

We will use the following `import` statement:

```
import QtMultimedia 5.12
```

Here's a simple stanza that will play the .mp3 file named `sample.mp3`:

```
Audio {
    id: audioPlayer
    source: "sample.mp3"
}
```

The `source` property is where you declare which file to play. Now, you just have to call the `play()` method to have this `sample.wav` file play:

```
Component.onCompleted: audioPlayer.play()
```

You can also set the `autoPlay` property to `true` instead of calling `play`, and that would play the file once the component is completed.

Setting the volume is as easy as declaring the `volume` property and setting a decimal value between 0 and 1—1 being full volume and 0 being muted:

```
volume: .75
```

> Getting the metadata or ID tags from the file is not obvious, as they only become available after the `metaDataChanged` signal gets emitted. This is only emitted by the `Audio` element's `metaData` object.

Sometimes, you might need to display a file's metadata, or the extra data that can be within the audio file's headers. The `Audio` component has a `metaData` property that can be used like this:

```
metaData {
    onMetaDataChanged: {
        titleLabel.text = "Title: " + metaData.title
        artistLabel.text = "Artist: " + metaData.contributingArtist
        albumLabel.text = "Album: " + metaData.albumTitle
    }
}
```

If you need to access the microphone and record audio, you will need to dive into C++, so let's take a look at `QAudioRecorder`.

QAudioRecorder

Recording audio is one of my passions. Recording audio, or more specifically using the microphone, may require user permissions on some platforms.

The recording of audio, called taping back in my day, can be implemented by using the `QAudioRecorder` class. Recording properties are controlled by the `QAudioEncoderSettings` class, from which you can control the codec that's used, the channel count, the bit rate, and the sample rate. You can either explicitly set the bit rate and sample rate, or use the more generic `setQuality` function.

 The source code can be found in this book's Git repository under the `Chapter09-2` directory, in the `cp9` branch.

You might want to query the input devices and see which settings are available. To do that, you would query using `QAudioDeviceInfo`, iterating through `QAudioDeviceInfo::availableDevices(QAudio::AudioInput)`:

```
void MainWindow::listAudioDevices()
{
    for (const QAudioDeviceInfo &deviceInfo :
        QAudioDeviceInfo::availableDevices(QAudio::AudioInput)) {
        ui->textEdit->insertPlainText(
                QString("Device name: %1\n")
                .arg(deviceInfo.deviceName()));

        ui->textEdit->insertPlainText(
                "    Supported Codecs: "
                + deviceInfo.supportedCodecs()
                .join(", ") + "\n");
        ui->textEdit->insertPlainText(
                QString("    Supported channel count: %1\n")
.arg(stringifyIntList(deviceInfo.supportedChannelCounts())));
        ui->textEdit->insertPlainText(
                QString("    Supported bit depth b/s: %1\n")
.arg(stringifyIntList(deviceInfo.supportedSampleSizes())));
        ui->textEdit->insertPlainText(
                QString("    Supported sample rates Hz: %1\n")
.arg(stringifyIntList(deviceInfo.supportedSampleRates())));
    }
}
```

Qt Multimedia uses the term sample sizes for the more common term bit depth.

As you can see from my laptop, I have a few different audio input devices. The laptop's built-in audio chip got fried from an electrical spike, which is why it isn't seen here:

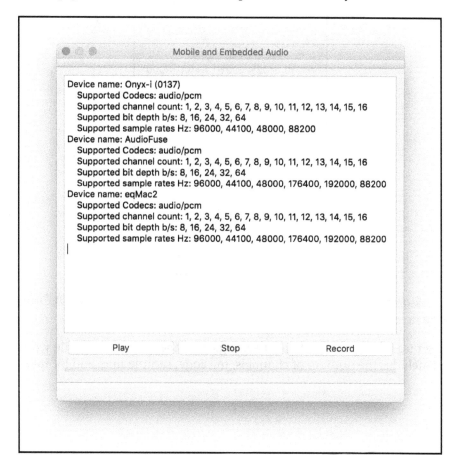

For iPhone, it is different. It has only one audio device, named `default`:

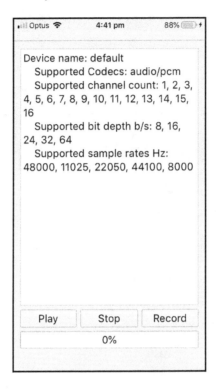

My Linux desktop reports a lot of audio input devices because of the ALSA Driver, which I won't include here.

We need to set up the recording encoder settings with the type of audio file we want to record. This includes the number of channels, the code, sample rate, and bit rate:

```
QAudioEncoderSettings audioSettings;
audioSettings.setCodec("audio/pcm");
audioSettings.setChannelCount(2);
audioSettings.setBitRate(16);
audioSettings.setSampleRate(44100);
```

If you want to let the system decide on the various settings, it is quicker and takes less code to use the `setQuality` function, which can take one of the following values:

- `QMultimedia::VeryLowQuality`
- `QMultimedia::LowQuality`
- `QMultimedia::NormalQuality`
- `QMultimedia::HighQuality`
- `QMultimedia::VeryHighQuality`

Let's choose `NormalQuality`, which will give the same results:

```
audioSettings.setQuality(QMultimedia::NormalQuality);
```

The `QAudioRecorder` class is used to record the audio, so let's construct a `QAudioRecorder` and set the encoding settings:

```
QAudioRecorder *audioRecorder = new QAudioRecorder(this);
audioRecorder->setEncodingSettings(audioSettings);
```

You can also specify which audio input to use, but first you will need to get a list of available audio input:

```
QStringList inputs = audioRecorder->audioInputs();
```

If you don't want to bother about which audio device to use, you can specify it using the default device with the `defaultAudioInput()` function:

```
audioRecorder->setAudioInput(audioRecorder->defaultAudioInput());
```

We can save it to a file, or even a network location, as the `setOutputLocation` function takes a `QUrl`. We will just specify a local file to save it to:

```
audioRecorder->setOutputLocation(QUrl::fromLocalFile("record1.wav"));
```

If the file is relative, like it is here, you can get the actual output location using `outputLocation()` once the recording has started.

Finally, we can start the recording process:

```
audioRecorder->record();
```

There are also the `stop()` and `pause()` methods to control the recording operation.

Of course, you will want to connect to the error signal, because errors can and will happen from time to time. Again, note the use of the `QOverload` syntax that's used in error-reporting signals:

```
connect(audioRecorder,
QOverload<QMediaRecorder::Error>::of(&QMediaRecorder::error),
            [=](QMediaRecorder::Error error){
                ui->textEdit->insertPlainText("QAudioRecorder Error: " +
audioRecorder->errorString());
                on_stopButton_clicked();
            });
```

So, now that we have recorded some audio, we might want to listen to it. This is where `QMediaPlayer` comes in.

QMediaPlayer

`QMediaPlayer` is fairly straightforward. It can play both audio and video, but here we will only be playing audio. First, we need to set up the media to play by calling `setMedia`.

We can use `QAudioRecorder` to get the output file and use it to play:

```
player = new QMediaPlayer(this);
player->setMedia(audioRecorder->outputLocation());
```

We will have to monitor the current playing position, so we will connect the `positionChanged` signal to a progress bar:

```
connect(player, &QMediaPlayer::positionChanged,
        this, &MainWindow::positionChanged);
```

Connect the error signal and its `QOverload` syntax:

```
connect(player, QOverload<QMediaPlayer::Error>::of(&QMediaPlayer::error),
            [=](QMediaPlayer::Error error){
            ui->textEdit->insertPlainText("QMediaPlayer Error: " +
player->errorString());
            on_stopButton_clicked();
    });
```

Then, just call `play()` on the `QMediaPlayer` object:

```
player->play();
```

You can even set the playback volume:

```
player->setVolume(75);
```

If you need access to the media data, let's say for getting the volume level of the data as it plays, you will want to use something other than QMediaPlayer to play your file.

QAudioOutput

QAudioOutput provides a way to send audio to an audio output device:

```
QAudioOutput *audio;
```

Using QAudioOutput, you will need to set up the exact format of your file. To get the format of your file, you could use QMediaResource.

Scratch that—QMediaResource is being depreciated in Qt 6.0, and does not do what the docs say it is supposed to do, and doesn't work like it should. We need to hardcode the data format, so we will use the basic good-quality stereo format. QAudioFormat is the way to do this:

```
QAudioFormat format;
format.setSampleRate(44100);
format.setChannelCount(2);
format.setSampleSize(16);
format.setCodec("audio/pcm");
format.setByteOrder(QAudioFormat::LittleEndian);
format.setSampleType(QAudioFormat::UnSignedInt);
```

We will iterate through the audio devices and check that QAudioDeviceInfo supports this format:

```
    for (const QAudioDeviceInfo &deviceInfo :
QAudioDeviceInfo::availableDevices(QAudio::AudioOutput)) {
        if (deviceInfo.isFormatSupported(format)) {
            audio = new QAudioOutput(deviceInfo, format, this);
            connect(audio, &QAudioOutput::stateChanged, [=] (QAudio::State
state) {
            qDebug() << Q_FUNC_INFO << "state" << state;
            if (state == QAudio::StoppedState) {
                if (audio->error() != QAudio::NoError) {
                    qDebug() << Q_FUNC_INFO << audio->error();
                }
            }
        }
```

```
        });
    }
```

Here, I connected to the `stateChanged` signal and tested whether the state is `StoppedState`; we know there might be an error, so we check the `error()` of the `QAudioOutput` object. Otherwise, we can play the file:

```
QFile sourceFile;
sourceFile.setFileName(file);
sourceFile.open(QIODevice::ReadOnly);
audio->start(&sourceFile);
```

We see now that Qt Multimedia has various ways of playing audio. Now, let's take a look at the camera and recording video.

Image sensors – camera

First, we should establish whether the device has any cameras. This helps us determine specifics about the use of the camera and other camera specifications, such as the orientation or position on the device.

For this, we will use `QCameraInfo`.

QCameraInfo

We can get a list of cameras using the `QCameraInfo::availableCameras()` function:

 The source code can be found in this book's Git repository under the `Chapter09-4` directory, in the `cp9` branch.

```
QList<QCameraInfo> cameras = QCameraInfo::availableCameras();
foreach (const QCameraInfo &cameraInfo, cameras)
    ui->textEdit->insertPlainText(cameraInfo.deviceName() + "\n");
```

On my Android device, I see two cameras, named back and front. You can also check for front and back cameras using QCameraInfo::position(), which will return one of the following:

- QCamera::UnspecifiedPosition
- QCamera::BackFace
- QCamera::FrontFace

FrontFace means that the camera lens is on the same side as the screen. You can then use QCameraInfo to construct a QCamera object:

```
QCamera *camera;
if (cameraInfo.position() == QCamera::BackFace) {
    camera = new QCamera(cameraInfo);
}
```

Now, check for the capture modes the camera supports, which can be one of the following:

- QCamera::CaptureViewfinder
- QCamera::CaptureStillImage
- QCamera::CaptureVideo

Let's do a quick still image shot first. We need to tell the camera to use the QCamera::CaptureStillImage mode:

```
camera->setCaptureMode(QCamera::CaptureStillImage);
```

The statusChanged signal is used to monitor the status, which can be one of the following values:

- QCamera::UnavailableStatus
- QCamera::UnloadedStatus
- QCamera::UnloadingStatus
- QCamera::LoadingStatus
- QCamera::LoadedStatus
- QCamera::StandbyStatus
- QCamera::StartingStatus
- QCamera::StoppingStatus
- QCamera::ActiveStatus

Let's connect to the `statusChanged` signal so that we can see status changes:

```
connect(camera, &QCamera::statusChanged, [=] (QCamera::Status status) {
    ui->textEdit->insertPlainText(QString("Status changed %1").arg(status)
+ "\n");
});
```

If you need to fiddle with any of the camera settings, you will have to `load()` it before you can get access to the `QCameraImageProcessing` object:

```
camera->load();
QCameraImageProcessing *imageProcessor = camera->imageProcessing();
```

With the `QCameraImageProcessing` class, you can set configurations, such as brightness, contrast, saturation, and sharpening.

Before we call start on the camera, we need to set up a `QMediaRecorder` object for the camera. Since `QCamera` is inherited from `QMediaObject`, we can feed it to the `QMediaRecorder` object.

 Qt Multimedia Widgets are not supported on Android.

I tried `QCamera` version 5.12 on both Mac and iOS, but it kept crashing when I tried to `start()` the camera. I was successful on Linux desktop. On Android, since multimedia widgets are not supported, the camera viewfinder widget did not work, but I could still capture images from the image sensor.

Maybe you'll have better luck with the QML side of things. QML APIs are usually optimized for easy use.

Camera

Yes, the QML `Camera` is so much easier to implement. Really, there are only two components you need to take a photo: `Camera` and `VideoOutput`.

`VideoOutput` is the element to use for the viewfinder. It is also used when you are recording video:

```
Camera {
    id: camera
    position: Camera.BackFace
    onCameraStateChanged: console.log(cameraState)
    imageCapture {
        onImageCaptured: {
            console.log("Image captured")
        }
    }
}
```

The `position` property controls which camera to use, especially on a mobile device that may have a front-facing and rear-facing camera. Here, I am not only using the rear camera. You would use the `FrontFace` position to take a selfie.

`imageCaptured` pertains to the `CameraCapture` sub-element. We can handle the `onImageCaptured` signal to preview the image or to alert the user that a photo has been taken.

The other properties of the `Camera` object can be controlled by their corresponding components:

- `focus : CameraFocus`
- `flash : CameraFlash`
- `exposure : CameraExposure`
- `imageProcessing : CameraImageProcessing`
- `imageCapture : CameraCapture`
- `videoRecorder: CameraRecorder`

`CameraRecorder` is what you would use to controls saturation, brightness, color filters, contrast, and other settings.

`CameraExposure` controls things such as aperture, exposure compensation, and shutter speed.

`CameraFlash` can turn the flash on, off, or use auto mode. It can also set red-eye compensation and video (constant) mode.

We need a view finder to see what the heck we are trying to capture, so let's take a look at the VideoOutput element.

VideoOutput

VideoOutput is the component we use to view what the camera is sensing.

 The source code can be found on the Git repository under the Chapter09-5 directory, in the cp9 branch.

To implement the VideoOutput component, you need to define the source property. Here, we are using the camera:

```
VideoOutput {
    id: viewfinder
    source: camera
    autoOrientation: true
}
```

The autoOrientation property is used to allow the VideoOutput component to compensate for the device orientation of the image sensor. Without this being true, the image might show up in the view finder with the wrong orientation and confuse the user, making it harder to take a good photo or video.

Let's make this VideoOutput clickable by adding a MouseArea, where I will use the onClicked and onPressAndHold signals to focus and actually capture an image:

```
MouseArea {
    anchors.fill: parent
    onPressAndHold: {
        captureMode: captureSwitch.position === 0 ?Camera.CaptureStillImage
: Camera.CaptureVideo
        camera.imageCapture.capture()
    }
    onClicked: {
        if (camera.lockStatus == Camera.Unlocked)
            camera.unlock();
            camera.searchAndLock();
    }
}
```

I also added a `Switch` component from Qt Quick Controls to control whether the user wants a still photo or video recorded.

To focus the camera, call the `searchAndLock()` method, which starts focus, white balance, and exposure computations.

Let's add support for recording videos. We will add a `CameraRecorder` container to the `Camera` component:

```
VideoRecorder {
    audioEncodingMode: CameraRecorder.ConstantBitrateEncoding;
    audioBitRate: 128000
    mediaContainer: "mp4"
}
```

We can set certain aspects for the video, such as bit rate, frame rate, number of audio channels, and what container to use.

We need to also change the way our `onPressAndHold` signal works to make sure we record video when the user has specified it, by the use of the switch:

```
onPressAndHold: {
    captureMode: captureSwitch.position === 0 ? Camera.CaptureStillImage :
Camera.CaptureVideo
    if (captureSwitch.position === 0)
        camera.imageCapture.capture()
    else
        camera.videoRecorder.record()
}
```

We need some way to stop recording, so let's modify the `onClicked` signal handler to stop the recording when it is in `RecordingState`:

```
onClicked: {
    if (camera.videoRecorder.recorderState ===
CameraRecorder.RecordingState) {
        camera.videoRecorder.stop()
    } else {
        if (camera.lockStatus == Camera.Unlocked)
            camera.unlock();
        camera.searchAndLock();
    }
}
```

Now, we need to actually see the video we just recorded. Let's move on and look at how to play a video.

Visual media – playing video

Playing a video with QML is much like playing audio using `MediaPlayer`, only using a `VideoOutput` instead of an `AudioOutput` component.

 The source code can be found on the Git repository under the `Chapter09-6` directory, in the `cp9` branch.

We begin by implementing a `MediaPlayer` component:

```
MediaPlayer {
    id: player
```

The property named `autoPlay` will control the automatic starting of the video once the component is completed.

Here, the `source` property is set to the filename of our video:

```
        autoPlay: true
        source: "hellowindow.m4v"
        onStatusChanged: console.log("Status " + status)
        onError: console.log("Error: " + errorString)
    }
```

We then create a `VideoOutput` component, with the source being our `MediaPlayer`:

```
VideoOutput {
    source: player
    anchors.fill : parent
  }

MouseArea {
    id: playArea
    anchors.fill: parent
    onPressed: player.play();
}
```

The `MouseArea`, which is the entire application, is used here to start playing the video when you click anywhere on the application.

With C++, you would use the `QMediaPlayer` class with a `QGraphicsVideoItem`, `QVideoWidget`, or something else.

Since `QMultimediaWidgets` have limited support on mobile devices, I will leave this as an exercise for the reader.

Qt Multimedia also supports FM, AM, and some other radios, providing your device has a radio in it as well.

Tuning it in – FM radio tuner

Some Android phones have an FM radio receiver. Mine does! It requires the wired headphones to be inserted to work as the antenna.

We start by implementing a `Radio` component:

```
Radio {
    id: radio
```

The `Radio` element has a `band` property that you can use to configure the radio's frequency band use. They are one of the following:

- `Radio.AM` : 520 - 1610 kHz
- `Radio.FM` : 87.5 - 108 MHz, Japan 76 - 90 MHz
- `Radio.SW` : 1.711 to 30 MHz
- `Radio.LW` : 148.5 to 283.5 kHz
- `Radio.FM2` : Range not defined

```
band: Radio.FM
Component.onCompleted {
    if (radio.availability == Radio.Available)
        console.log("Good to go!")
    else
        console.log("Sad face. No radio found. :(")
    }
}
```

The `availability` property can return the following different values:

- `Radio.Available`
- `Radio.Busy`
- `Radio.Unavailable`
- `Radio.ResourceMissing`

The first thing the user will do with a radio is scan for stations, which can be accomplished by using the `searchAllStations` method, which takes one of the following values:

- `Radio.SearchFast`
- `Radio.SearchGetStationId`: Like `SearchFast`, it emits the `stationFound` signal

The `stationsFound` signal returns an `int frequency` and `stationId` string for each station that's tuned in. You could collect these in a model-based component, such as `ListView`, using a `ListModel`. The `ListView` would use the `ListModel` as its model.

You can cancel the scan by calling the `cancelScan()` method. The `scanUp()` and `scanDown()` methods are similar to `searchAllStations`, except it does not remember the stations it found. The `tuneUp` and `tuneDown` methods will tune the frequency up or down one step, according to the `frequencyStep`.

Here are some other interesting properties:

- `antennaConnected`: True if an antenna is connected
- `signalStrength`: Strength of the signal in %
- `frequency`: Holds and sets the frequency that the radio is tuned to

Summary

In this chapter, we discussed the different aspects of the big API of Qt Multimedia. You should now be able to position sound in a 3-dimensional way for 3D games. We learned how to record and play audio and video, and control and use the camera to take a selfie. We also touched on using QML to listen to radio stations.

In the next chapter, we will dig into using `QSqlDatabase` to access databases.

10
Remote Databases with Qt SQL

Qt SQL is not dependent on any particular database driver. The same API can be used with various popular database backends. Databases can get huge storage, whereas mobile and embedded devices have limited amounts of storage, more so with embedded devices than mobile phones. You will learn about using Qt to access databases remotely over the network.

We will cover the following topics in this chapter:

- Drivers
- Connecting to database
- Creating a database
- Adding to a database

Technical requirements

You can grab this chapter's source code in the `cp10` branch at `git clone -b cp10 https://github.com/PacktPublishing/Hands-On-Mobile-and-Embedded-Development-with-Qt-5`.

You should also have installed the `sqlite` or `mysql` package for your system.

Drivers are database backends

Qt supports a variety of database drivers or backends to the databases. The backends wrap the various system databases and allow Qt to have a unified API frontend. Qt supports the following database types:

Database types	Software
QDB2	IBM Db2
QIBASE	Borland InterBase
QMYSQL	MySQL
QOCI	Oracle Call Interface
QODBC	ODBC
QPSQL	PostgreSQL
QSQLITE	SQLite version 3 or above
QSQLITE2	SQLite version 2
QTDS	Sybase Adaptive Server

We will be looking into QMYSQL type, since it supports remote access. MySQL can be installed on Raspberry Pi. QSQLITE3 can be shared on a network resource and made to support remote access, and iOS and Android have support for SQLite.

Setup

The MySQL database will need to be configured to let you have remote access to it. Let's look at how we can do this:

1. You will need to have the server and/or client installed.
2. Then, we'll create the database and make it accessible from the network, if needed. This will be done using the command line and a Terminal application.

The MySQL server

I am using Ubuntu, so these commands will be mostly specific to a Debian-based Linux. If you are using a different Linux distribution, only the installation command would be different. You should install the MySQL server and client according to your distribution. The commands to create the database would be the same.

Here's how we will set up the server:

1. You will need the MySQL server and client installed:

   ```
   sudo apt-get install mysql-server mysql-client
   ```

2. Run `sudo mysql_secure_installation`, which will allow you to set up the
   root account. Then, log in to the `mysql` root account:

   ```
   sudo mysql -u root -p
   ```

3. Create a new database `username` and `password`: `GRANT ALL PRIVILEGES ON
   . TO 'username'@'localhost' IDENTIFIED BY 'password';`.
   Change `username` to your database user, and `password` to a password you want
   to use to access this database.

4. To make the server accessible from a host other than localhost,
   edit `/etc/mysql/mysql.conf.d/mysqld.cnf`.

5. Change the `bind-address = localhost` to `bind-address = <your ip>`
   line, `<your ip>` being the IP address of the machine that the database is on.
   Then, restart the `mysql` server:

   ```
   sudo  service mysql restart
   ```

Back in your MySQL console, let a remote user access the database:

```
GRANT ALL ON *.* TO 'username'@'<your ip>' IDENTIFIED BY 'password';
```

Change `<your ip>` to the IP address or hostname of the client device, `username` to the
username you used on the MySQL server, and `password` to the password they will use.

SQLite

SQLite is a file-based database and, as such, there is no such thing as a server. We can still
connect to it remotely via a network filesystem, such as Windows file sharing/Samba,
Network File System (NFS), or the **Secure Shell File System** (**SSHFS**) on Linux. SSHFS
allows you to mount and access a remote filesystem like it is a local filesystem.

There's no need to create a database manually using arcane commands unless you need to,
as we will create it using Qt!

On Android, there are Samba clients, which will mount a Windows network share so we can use that. If you are using a Raspberry Pi or some other development board, you might be able to use SSHFS to mount a remote directory over SSH.

Connecting to a local or remote database

Once we have the database configured and running, we can now connect to it using the same functions regardless of whether it is local or a remote database. Now, let's take a look at writing code to connect to a database, whether local or remote.

Databases are either locally available, which usually means on the same machine, or accessed remotely over a network. Connecting to these different databases using Qt is essentially the same. Not all databases support remote access.

Let's begin by using a local database.

To use the `sql` module, we need to add `sql` to the profile:

```
QT += sql
```

To connect to a database in Qt, we need to use the `QSqlDatabase` class.

QSqlDatabase

Despite the name, `QSqlDatabase` represents a connection to a database, not the database itself.

To create a connection to a database, you first need to specify which database type you are using. It is referenced as a string representation of the supported database. Let's first choose the MySQL `QMYSQL` database.

 The source code can be found on the Git repository under the `Chapter10-1` directory, in the `cp10` branch.

To use `QSqlDatabase`, we first need to add the database to create its instance.

The static `QSqlDatbase::addDatabase` function takes one parameter, that of a database type, and adds the instance of the database to the list of database connections.

Here, we add a MySQL database, so use the QMYSQL type:

```
QSqlDatabase db = QSqlDatabase::addDatabase("QMYSQL");
```

If you are connecting to a SQLite database, use the MSQLITE database type:

```
QSqlDatabase db = QSqlDatabase::addDatabase("MSQLITE");
```

Most databases require a username and password. To set the username and password, use the following:

```
db.setUserName("username");
db.setPassword("password");
```

Since we are connecting to a remote MySQL database, we need to specify the hostname as well. It can also be an IP address:

```
db.setHostName("10.0.0.243");
```

To start the connection, call the open() function:

```
bool ok = db.open()
```

open returns a bool, which is true if it was successful, or false if it failed, in which case we can check the error:

```
if (!db.open()) {
    qWarning() << dq.lastError.text();
}
```

If this opens successfully, we are connected to the database.

Let's actually create the remote database, since we have the needed permissions.

Creating and opening a database

For the SQLite database, once we open it, it creates the database on the filesystem. For MySQL, we have to send MySQL commands to create the database. We construct the SQL query using QSqlQuery do this in MySQL. QSqlQuery takes the database object as an argument:

```
QSqlQuery query(db);
```

To send a query, we call the `exec()` function on the `QSqlQuery` object. It takes a `String` as a typical `query` syntax:

```
QString dbname = "MAEPQTdb";
if (!query.exec("CREATE DATABASE IF NOT EXISTS " + dbname)) {
    qWarning() << "Database query error" << query.lastError().text();
}
```

`dbname` here is any `String` we want the database name to be; I am using `MAEPQT db`.

If this command fails, we issue a warning message. If it succeeds, then we go on and issue the command to `USE` it, so we call another `query` command:

```
query.exec("USE " + dbname);
```

From here, we need to create some tables. I will keep it simple and fill it with some data.

We start another query, but with an empty command, and the `db` object as the second argument, which will create the `QSqlQuery` object on the specified database, but does not execute any commands until we are ready:

```
QSqlQuery q("", db);
q.exec("drop table Mobile");
q.exec("drop table Embedded");
q.exec("create table Mobile (id integer primary key, Device varchar, 
Model varchar, Version number)");

q.exec("create table Embedded (id integer primary key, Device varchar,Model 
varchar, Version number)");
```

The database is prepared, so now we can add some data.

Adding data to the database

The Qt documents state that it is not a best practice to keep the `QSqlDatabase` object around.

Here are a few different ways we could go about this:

1. We could use `QSqlDatabase::database` to grab an instance of the opened database:

```
QSqlDatabase db = QSqlDatabase::database("MAEPQTdb");
QSqlQuery q("", db);
q.exec("insert into Mobile values (0, 'iPhone', '6SE', '12.1.2')");
q.exec("insert into Mobile values (1, 'Moto', 'XT1710-09', '2')");
q.exec("insert into Mobile values (1, 'rpi', '1', '1')");
q.exec("insert into Mobile values (1, 'rpi', '2', '2')");
q.exec("insert into Mobile values (1, 'rpi', '3', '3')");
```

2. We can also use another function of `QSqlQuery`, named `prepare()`, which prepares the query string for execution using a proxy variable.

   Then, we can use `bindValue` to bind the value to its identifier:

```
q.prepare("insert into Mobile values (id,   device, model, version)"
        "values ( :id, :device, :model, :version)");

q.bindValue(":id", 0);
q.bindValue(":device", "iPhone");
q.bindValue(":model", "6SE");
q.bindValue(":version", "12.1.2");
q.exec();
```

3. As an alternative, you can call `bindValue` with the first argument being the index of the position of the identifier, starting at the number 0 and working upward through the values:

```
q.bindValue(1, "iPhone");
q.bindValue(3, "12.1.2");
q.bindValue(2, "6SE");
```

4. You can also use `bindValue` in the order of values:

```
q.bindValue(0);
q.bindValue("iPhone");
q.bindValue("6SE");
q.bindValue("12.1.2");
```

Next, let's look at retrieving data from the database.

Executing queries

So far, we have been running queries, but not getting any data in return. One of the points of a database is to query for data, not just to enter it. What fun would it be if we could only input data? The Qt API has a way to accommodate the different syntax and millions of ways a query can be made. Most of the time, it is specific to the type of data you need to be returned, but also specific to the database data itself. Luckily, QSqlQuery is general enough that the query parameter is a string.

QSqlQuery

To retrieve data, execute a query using QsqlQuery and then operate on the records using the following functions:

- first()
- last()
- next()
- previous()
- seek(int)

The first() and last() functions will retrieve the first and last records respectively. To iterate backward through the records, use previous(). The seek (int) function takes on integer as an argument to determine which record to retrieve.

We will use next(), which will iterate forward through the records found in the query:

```
QSqlDatabase db = QSqlDatabase::database("MAEPQTdb");
QSqlQuery query("SELECT * FROM Mobile", db);
int rowCount = 0;
while (query.next()) {
    QString id = query.value(0).toString();
    QString device = query.value(1).toString();
    QString model = query.value(2).toString();
    QString version = query.value(3).toString();
    ui->tableWidget->setRowCount(rowCount + 1);
    ui->tableWidget->setItem(rowCount, 0, new QTableWidgetItem(id));
    ui->tableWidget->setItem(rowCount, 1, new
QTableWidgetItem(device));
    ui->tableWidget->setItem(rowCount, 2, new
QTableWidgetItem(model));
    ui->tableWidget->setItem(rowCount, 3, new
QTableWidgetItem(version));
```

```
        rowCount++;
    }
```

We also use `value` to retrieve each field's data, which takes an `int` that indicates the position of the record starting at 0.

You can also use `QSqlRecord` and `QSqlField` to do the same thing, but with more clarity as to what is actually going on:

```
QSqlField idField = record.field("id");
QSqlField deviceField = record.field("device");
QSqlField modelField = record.field("model");
QSqlField versionField = record.field("version");
qDebug() << Q_FUNC_INFO
        << modelField.name()
        << modelField.tableName()
        << modelField.value();
```

To get the record data, use `value()`, which will return a `QVariant` that represents the data for that record field.

We could have used a model-based widget and then used `QsqlQueryModel` to execute the query.

QSqlQueryModel

`QSqlQueryModel` inherits from `QSqlQuery`, and returns a model object that can be used with model-based widgets and other classes. If we change our `QTableWidget` to `QTableView`, we can use `QSqlQueryModel` to be its data model:

```
QSqlQueryModel *model = new QSqlQueryModel;
model->setQuery("SELECT * FROM Mobile", db);
tableView->setModel(model);
```

Here is my Raspberry Pi (with a Lego stand!) running the database example using the MySQL plugin remotely:

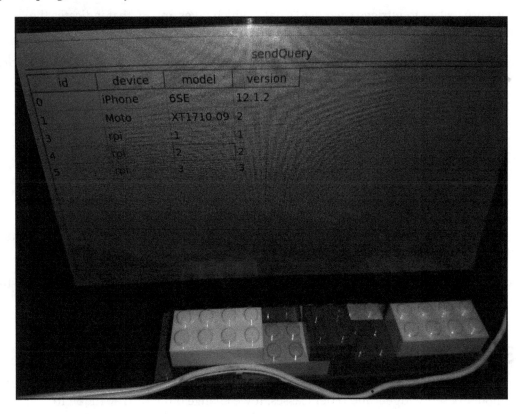

Summary

In this chapter, we learned that QSqlDatabase represents a connection to any supported database. You can use this to log in to remote MySQL databases or a SQLite database on a network share. To perform data retrieval, use QSqlQuery. You use the same class to set data and tables and other database commands. You can use QSqlDatabaseModel if you are using a model-view application.

In the next chapter, we will explore in-app purchasing using the Qt Purchasing module.

Enabling In-App Purchases with Qt Purchasing

11

In-app purchasing on mobile phones is essential to generate more income. Qt utilizes system APIs to bring in-app purchases to Qt apps. Android and iOS both have their own app stores, and each store has its own methods for registering products for sale. This is where Qt Purchasing comes in!

In this chapter, we will cover the following topics:

- Registering in Android and iOS stores
- Creating an in-app product
- Restoring purchases

Registering on Android Google Play

Selling mobile applications is often hit-and-miss; only a few apps that are available to buy actually make money. One of the best ways to make money at the moment is to make your application free to download, but to include in-app purchases. That way, people get to try out your app but also have the opportunity to make purchases if they want enhanced play. This section presumes you have already registered your app to the relevant mobile store. To activate in-app purchases, you first need to register the things you intend to sell. This has its own benefits, as it allows testers to *buy* and install things you intend to sell.

Let's take a look at how to do this on Android devices first:

1. You will first need to register for a Google Developers account in order to create an application that will be available in Google Play Store, `https://developer.android.com`.

2. You will also need to add and edit the `AndroidManifest.xml` file.

   In Qt Creator, navigate to:
   Projects | Build | Build Settings | Build Android APK | Create Templates. Here, you will need to edit the **Package** name, ideally using the convention, `com.<company>.<application name>`. Other naming conventions are of course available, as you can name it anything you want.

3. The version number must be incremented when you update your package in the Google Play Store. The easiest way to do this is to tick the box labeled **Include default permissions for Qt Modules**. If not, you need to be sure to add the `uses-permission android:name="com.android.vending.BILLING"` permission.

4. You will also need to sign this package with your certificates, so create a keystore if you haven't already done this.

5. In-app purchasing in Google is named **Google Play Billing**, while **Google Play Console** is the name of the website you go through to publish apps to the Google Play Store. You need to register as a developer and pay a registration fee. (For me, it was 25 AUD.) Once the fee is paid, you can set up a merchant account.

6. After that, it's time to supply information about your application and upload store graphics such as screenshots and promotional videos. This is where you also need to supply contact details for your customers.

7. You can develop in-app purchasing by making an internal test available only to developers in your organization. Once you have the kinks worked out, and your app enters the alpha stage, you can broaden your test and make a closed test. After that, during beta development, you can have an open test.

8. On the Google Play Console website, click on your application and navigate to **Store presence | In-app products | Managed Products.**

9. Then, click on the blue button labelled **CREATE MANAGED PRODUCT**, as shown in the following screenshot:

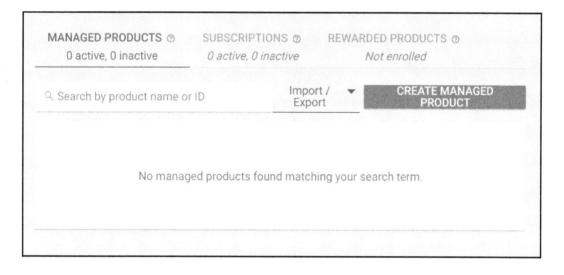

That will open a new form titled **New managed product**, as shown in the following screenshot:

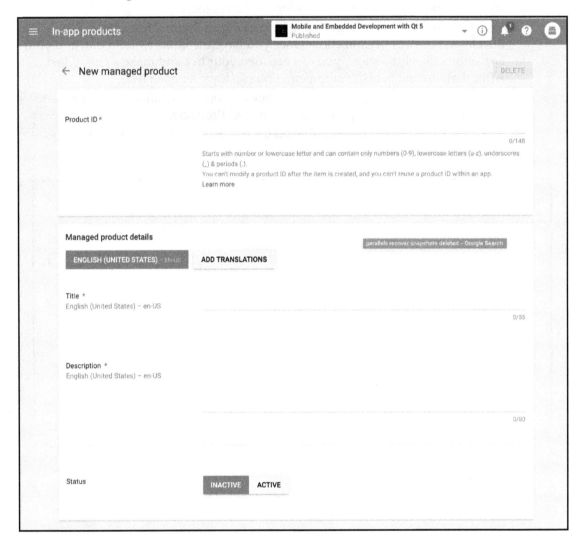

10. On this form, complete the following fields:

- **Product ID**: This will be used in the Qt app identifier
- **Title**
- **Description**
- **Status**: **ACTIVE** or **INACTIVE**
- **Pricing**: This is limited to be between $0.99 and $550.00

11. Then, click Save. You will be registered on Google Play.

> If you use the same **Product ID** for both Android and iOS, it will make the process of developing in-app purchases easier.

Registering on iOS App Store

You should already be enrolled on the Apple Developer Program and to have accepted all of the necessary agreements relating to tax, banking, and other data.

This section assumes you have already registered an app ID, have signed the relevant agreements, and so on. Registering an in-app purchase on iOS is fairly straightforward:

1. Navigate to your **Apple App Store Connect** account and sign in. Click on **Apps**, as we will be registering an application's in-app products.

2. Click on your app and then select **Features**. At the top of the page, click on the blue circle that contains a plus sign that is labelled **In-App Purchases (0)**, as shown in the following screenshot:

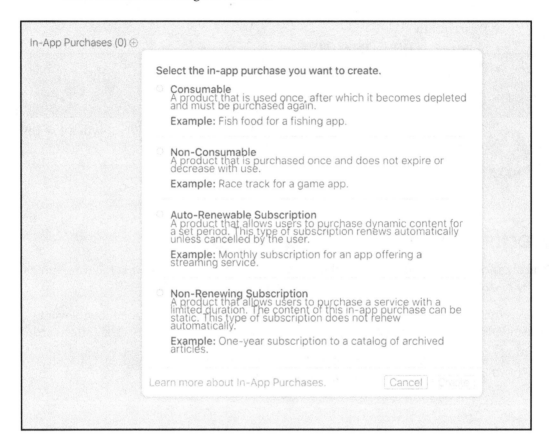

You can choose from the following options:

Consumable	Items that are used once by the app and need to be re-purchased
Non-Consumable	Items that do not expire but are purchased once
Auto-Renewing Subscription	Subscription content that is automatically renewed
Non-Renewing Subscription	Subscription content that is not renewed

3. You will have to fill in a form for this part of the process, so decide beforehand on the values for the following labelled items:

- **Reference name**
- **Product ID**
- **Price**
- **Tiered prices (start at $1.49)**
- **Start Date**
- **End Date**
- **Display Name**
- **Description**
- **Screenshot**
- **Review Notes**

Once you have the product ID, take note of this information for later. You will need it once you create your in-app purchase product with Qt Creator.

Creating an in-app product

Now, the real fun begins! Suppose you have designed a treasure-hunting game where users move around a map and look for treasure. In this scenario, you may want to offer accelerated gameplay, where users can purchase hints to help them to find the game's hidden treasure.

In our example, we will be selling colors. Colors are really great, as they are collectable and can be sold and traded by users with each other.

When you have developed and registered your app as was mentioned in the last sections on *Registering on iOS App Store* and *Registering on Android Google Play*, you can now develop and test Qt Purchasing. We will start by using QML.

The import line in your QML app to use Qt Purchasing is as follows:

```
import QtPurchasing 1.0
```

Add the following line to the profile:

```
QT += purchasing
```

Now, decide what your in-app purchase is going to be. Note that Qt Purchasing has the following two product types:

- Consumable
- Unlockable

Consumable purchases are things such as game tokens that are used once and can be purchased more than once. One example of this is game currency.

Unlockable purchases are features such as additional characters, advertisement removal, and level unlocking that can be re-downloaded, restored, or even transferred.

Our color product is a consumable purchase, enabling users to buy as many colors as they want.

In QtPurchasing, there are the following three QML components:

- `Product`
- `Store`
- `Transaction`

Store

The `Store` component represents the platform's default marketplace; on Android, it is the Google Play Store, and on iOS, it's the Apple App Store. A `Store` element has one method, `restorePurchases()`, which is used a user wants to restore their purchases.

You can either make `Product` a child of `Store`, or standalone, where the `Store` object is specified by an ID.

Product

The `Product` component represents an in-app purchase product. The `identifier` property corresponds to the product ID you used in the relevant stores when registering your in-app purchase products.

There are a few things to keep in mind about the `Product` component:

- `Product` can either be a child of `Store`, or it can be referred to by using the `id` property of `Store`
- `Product` can have one of two types: `Product.Consumable` or `Product.Unlockable`
- A `Product.Consumable` product can be purchased more than once, provided that the purchase has been finalized
- A `Product.Unlockable` product is purchased once and can be restored

The following code demonstrates a `Product` component that has the type of `Product.Consumable`:

```
Store {
    id: marketPlace
}

Product {
    id: colorProduct
    store:marketPlace
    identifier: "some.store.id"
    type: Product.Consumable
}

Button {
    text: "Buy this color"
    onClicked: {
        colorProduct.purchase()
    }
}
```

Now, it's time to move on to our purchase options. Take at look at the following screenshot:

To start the purchase procedure, use the `purchase()` method, which the **OK** button calls to bring up the following dialog from Google Play Store:

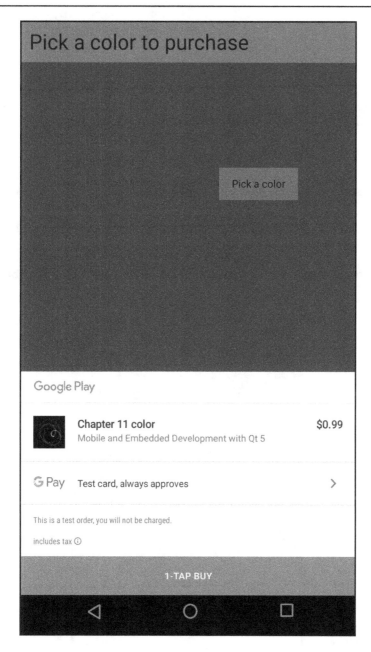

Notice that the payment made in the preceding screenshot is not a real one, but is instead made with the Google test card. No money was exchanged.

You will now want to handle the onPurchaseSucceeded and onPurchaseFailed signals. If you have products that can be restored, do so in the onPurchaseRestored signal, as follows:

```
onPurchaseSucceeded: {
    console.log("sale succeeeded " +  transaction.orderId)
// do something like fill a model view

    transaction.finalize()
}
```

You should also save the transaction. When the app starts up, it queries the Store of any purchases. If the user has purchased products, the onPurchaseSucceeded signal will get called with the transition ID for each purchase, so the app knows what purchases have already been made and can act accordingly.

The following screenshot illustrates a successful purchase on Google Play Store:

If the purchase fails for whatever reason, `onPurchaseFailed` will be called, as follows:

```
onPurchaseFailed: {
    console.log("product purchase failed " + transaction.errorString)
    transaction.finalize()
}
```

You may want to provide a user notification for either of the events we've looked at here, simply to provide clarity and avoid confusion for the user.

Transaction

`Transaction` represents the purchased product in the market store and contains properties regarding the purchase, including `status`, `orderId`, and a string describing any error that may have occurred. The following table explains these properties:

errorString	A platform-specific string that describes an error
failureReason	Can be either `NoFailure`, `CanceledByUser`, or `ErrorOccurred`
orderId	A unique ID issued by the platform store
product	The `product` object
status	Can be either `PurchaseApproved`, `PurchaseFailed`, or `PurchaseRestored`
timestamp	The time a transaction occurred

`Transaction` has one method: `finalize()`. All transactions need to be finalized whether they succeed or fail.

Once a user has successfully purchased a color, they should see something like the following screenshot:

Note that unlockable products can be restored. Let's move on and take a look at how that can be handled.

Restoring purchases

A user may want to restore purchases for a number of reasons. Perhaps they have re-installed the app, switched to a new phone, or even reset their existing phone.

 Only unlockable products can be restored.

Restoring purchases are initialized via the `restorePurchases()` method, which will then call the `onPurchaseRestored` signal for each purchase that gets restored, as follows:

```
Button {
    text: "Restore Purchases"
    onClicked: {
        colorProduct.restorePurchases()
    }
}
```

In the `product` component, it appears as follows:

```
onPurchaseRestored: {
    // handle product
    console.log("Product restored "+ transaction.orderId)
}
```

As you can see, QML makes it super simple to add in-app purchases and even to restore them if and when the need arises.

Summary

Qt makes it fairly simple to implement in-app purchases. Most of the work will involve getting your app together, and registering in-app products in your platform's store.

You should now be able to register an in-app purchase product with the relevant app stores. You should also know how to use QML to implement the in-app purchase product and make a store transition. We also explored how to restore any unlockable product purchases.

This chapter was all about mobile phone applications and purchases. In the next chapter, we will look at various cross-compiling methods and at how to debug remotely with an embedded device.

Section 4: Mobile Deployment and Device Creation

Compiling for, and deploying to, mobile and embedded devices can be challenging. In the case of embedded devices, readers may need to deploy to a barebones machine—operating system and all. The different methods to deploy cross-platform applications to mobile and embedded devices, as well as how to create a boot to Qt for a device using a Raspberry Pi, will be covered in this section.

This section comprises of the following chapters:

- Chapter 12, *Cross Compiling and Remote Debugging*
- Chapter 13, *Deploying to Mobile and Embedded*
- Chapter 14, *Universal Platform for Mobiles and Embedded Devices*
- Chapter 15, *Building a Linux System*

12
Cross Compiling and Remote Debugging

Since there is a good chance of using a Linux system on an embedded device, we will go through the steps needed to set up a cross compiler on Linux. Mobile phone platforms have their own ways of development, which will also be discussed. You will learn to compile cross-platforms for a different device and debug remotely via a network or USB connection. We will go through various mobile platforms.

We will cover the following topics in this section:

- Cross compiling
- Connecting to a remote device
- Debugging remotely

Cross compiling

Cross compiling is a method for building applications and libraries on a host machine for a different architecture than what is running on the host machine. When you build for a phone using an Android or iOS SDKs, you are cross compiling.

One easy way to do this is to use Qt for Device Creation, or Qt's Boot to Qt commercial tools. It is available for evaluation or purchase.

You do not have to build any of the tools and device system images yourself. I used Boot to Qt for my Raspberry Pi. This made set up a lot faster and easier. There are also more traditional ways of building for different devices, and they would be about the same, except for the target machine.

If you are on Windows, cross compiling can be a bit more tricky. You either install MinGW or Cygwin to build your own cross compiler, install Windows Subsystem for Linux, or install a prebuilt cross `toolchain`, for example, from Sysprogs.

Traditional cross tools

There are many ways to get a cross compiler. Device makers can release a cross `toolchain` with their software stack. Of course, if you are building your own hardware, or just want to create your own cross `toolchain`, there are other options. You can download a prebuilt cross `toolchain` for your device's architecture, or you can build it yourself. If you do end up compiling the `toolchain`, you will need a fast and robust machine with a lot of disk space, as it will take quite a long time to finish and use a lot of filesystem—easily 50 GB if you build the entire system.

DIY toolchain

There are also projects for which you can or must (if there is no supplied `toolchain`) build your own `toolchain`. The following are some of the more well known cross tools:

- **Buildroot**: `https://buildroot.org/`
- **Crosstool-NG**: `http://crosstool-ng.github.io/`
- **OpenEmbedded**: `http://www.openembedded.org`
- **Yocto**: `https://www.yoctoproject.org/`
- **Ångström**: `http://wp.angstrom-distribution.org/`

BitBake is used by OpenEmbedded, Yocto, and Ångström (as well as Boot to Qt), so it might be easiest to start out with one of those. You could say it is *Buildroot 2.0*, as it is the second incarnation of the original Buildroot. It is a completely different construction though. Buildroot is simpler and has no concept of packages, and thus, upgrading the system can be more difficult.

I will describe building a `toolchain` with BitBake in Chapter 15, *Building a Linux System*. Essentially it is very similar to building a system image; in fact, it has to build the `toolchain` before it can build the system image.

Buildroot

Buildroot is a tool that helps build complete systems. It can build the cross `toolchain` or use an external one. It traditionally uses an ncurses interface for configuration, much like the Linux kernel. It also has a new ncurses configurator, but also a Qt-based one. Let's use that!

In the directory where you unpacked Buildroot, run the following command:

```
make xconfig
```

Bah! It uses Qt 4. If you don't want to install Qt 4, you can always use `make menuconfig` or `make nconfig`.

Here is what the Qt interface looks like:

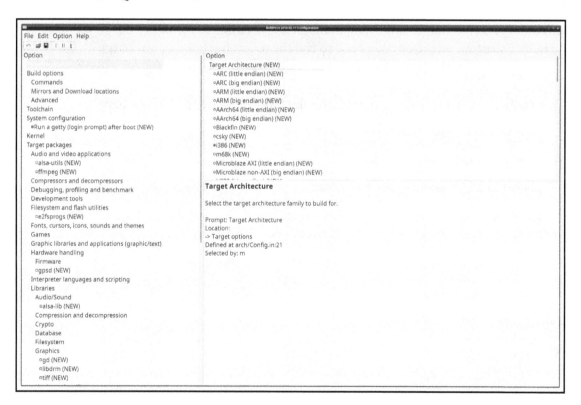

By default, Buildroot will create a system based on BusyBox, instead of glibc.

Once you have configured your system, save the configuration and close the configurator. Then run `make`, sit back, and let it build. It will place files into a directory called `output/`, under which your system image is in a directory named image.

Crosstool-NG

Crosstool-NG is meant for building toolchains, not system images. You can use the `toolchain` built with crosstools to build a system, although you would have to do it manually.

Crosstool-NG is similar to Buildroot, in that it uses ncurses to configure the `toolchain` to be built. Once you unpack it, you need to run the following `bootstrap` script:

```
./bootstrap
```

To install it, you would call configure with the following `--prefix` argument:

```
./configure --prefix=/path/to/output
```

You can also run it locally as follows:

```
./configure --enable-local
```

It will tell you any packages that are missing to install. On my Ubuntu Linux, I had to install `flex`, `lzip`, `help2man`, `libtool-bin`, and `ncurses-dev`.

Then run `make` and `make install` it you configured with a prefix.

You will need to add `/path/to/output/bin` into your `$PATH`

```
export PATH=$PATH:/path/to/output/bin.
```

Now you can run the following configuration:

```
./ct-ng menuconfig
```

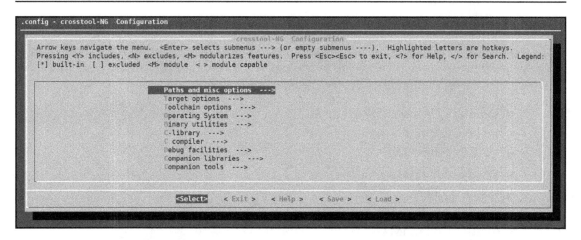

Then run `make`, which will build the cross `toolchain`.

Prebuilt tools

There are companies that make the following prebuilt cross tool chains for various devices and architectures:

- **Code Sourcery**: `http://www.codesourcery.com/`
- **Bootlin**: `https://toolchains.bootlin.com/toolchains.html`
- **Linaro, Debian, Fedora**: download from package manager
- **Boot to Qt**: `https://doc.qt.io/QtForDeviceCreation/qtb2-index.html`
- **Sysprogs**: `http://gnutoolchains.com/`

These are a few of the better ones. I have experienced the majority of these and used them at one time or another. Each comes with it's own instructions on how to install and use. Linaro, Debian, and Fedora all make ARM cross compilers. This is a book on Qt development, so I will describe Qt Company's offering—Boot to Qt.

Boot to Qt

Qt Company's Boot to Qt product comes complete with development tools and a prebuilt operating system image that you write to a micro SD card or flash to run on the device. They support the following other devices besides the Raspberry Pi:

- Boundary Devices i.MX6 Boards
- Intel NUC
- NVIDIA Jetson TX2
- NXP i.MX 8QMax LPDDR4
- Raspberry Pi 3
- Toradex Apalis iMX6 and iMX8
- Toradex Colibri iMX6, iMX6ULL, and iMX7
- WaRP7

I picked the RPI, as I already have a model 3 lying around with a touch screen.

When you run the system image, you boot into a Qt app that serves as a launcher for example apps. It also sets up Qt Creator to be able to run cross compiled apps on the device. You can run it on the device by hitting the Run button in Qt Creator.

Boot to Qt is a really fast and easy way to get a prototype up and running on a touch screen with a relatively small system. The Qt Company is currently working on getting Qt working well on smaller devices, such as microcontrollers.

You can run the Boot to Qt `toolchain` directly; you simply have to source the environment file. In the case of Raspberry Pi and Boot to Qt, it's called `environment-setup-cortexa7hf-neon-vfpv4-poky-linux-gnueabi`. You can also call the qmake of `toolchain` directly and run it on your profile `/path/to/x86_64-pokysdk-linux/usr/bin/qmake myApp.pro`.

A third option here is to just use Qt Creator and pick the Raspberry Pi as the target.

If you use Windows, there are a few options you can use to get a cross compiler `toolchain`.

Cross toolchains on Windows

There are a few ways you can cross compile on Windows, and we can briefly go through them. They are as follows, but there are undoubtedly others that are not covered here:

- Sysrogs provides prebuilt cross `toolchain` for use on Windows.
- Windows Subsystem for Linux.

Sysprogs

Sysprogs is a company that makes cross tool chains for targeting Linux devices that runs on Windows. Their `toolchain` can be downloaded from `http://gnutoolchains.com/`

1. Once installed, start a Qt 5.12.1 (MinGW 7.3.0 64-bit) console terminal
2. You need to add the `toolchain` to your path as follows:

       set PATH=C:\SysGCC\raspberry\bin;%PATH%

3. Add the `PATH` to Qt's `mingw` as follows:

       set PATH=C:\Qt\Tools\mingw730_64\bin;%PATH%

You will also have to build OpenGL and other requirements for Qt.

Configure Qt to cross compile as follows:

    ..\qtbase/configure -opengl es2 -device linux-rasp-pi-g++ -device-option
    CROSS_COMPILE=C:\SysGCC\raspberry\bin\arm-linux-gnueabihf- -sysroot
    C:\SysGCC\raspberry\arm-linux-gnueabihf\sysroot -prefix /usr/local/qt5pi -
    opensource -confirm-license -nomake examples -make libs -v -platform win32-
    g++

Windows Subsystem for Linux

You can install Windows Subsystem for Linux to install a cross compiler, which can be downloaded from `https://docs.microsoft.com/en-us/windows/wsl/install-win10`.

You can then pick the required Linux distribution—Ubuntu, OpenSUSE, or Debian. Once this is installed, you can use the built-in package manager to install the `toolchain` for Linux.

Mobile platform-specific tools

Both iOS and Android have prebuilt cross tools and SDKs that are available to download. You will need either of these if you are going to use Qt on the mobile platforms, as Qt Creator depends on the native platform build tools.

iOS

Xcode is the IDE beast you want to download, and it only runs on macOS X. You can get it from the App store on your desktop if you do not already have it. You will need to register as an iOS developer. From there, you can select the iOS build tools to download and set up. It's fairly automatic once you start the download.

You can also use these tools from the command line, but you need to install the command line tools from within Xcode. For Sierra, you can simply type the `gcc` command in the terminal. In that case, the system will open a dialog asking you if you want to install the command line tools. Alternatively, you can install it by running `xcode-select --install`.

I don't know of any tools for embedded systems that you can use with Xcode, unless you count the iWatch or iTV SDKs. Both of these you can download through Xcode.

You could use Darwin, of course, since it is open source and based on **Berkeley Software Distribution** (**BSD**). You could also use BSD. This is far from being able to run an Apple operating system on arbitrary embedded hardware, so your choices are limited.

Android

Android has Android Studio for its IDE development package and is available for macOS X, Windows, and Linux systems.

Like Xcode, Android Studio has command line tools as well, which you install through the SDK manager or the `sdkmanager` command.

`~/Android/Sdk/tools/bin/sdkmanager --list` will list all packages available. If you wanted to download the `adb` and `fastboot` commands, you could do the following:

`~/Android/Sdk/tools/bin/sdkmanager install "platform-tools"`

Android has catchy code names for their different versions, which is completely different from their API level. You should stick with the API level when installing Android SDKs. I have an Android phone that runs Android version 8.0.0, which has the code name Oreo. I would need to install an SDK for the API level 26 or 27. If I wanted to install the SDK, I might do the following:

```
~/Android/Sdk/tools/bin/sdkmanager install "platforms;android-26"
```

For working in Qt, you also need to install the Android NDK. I have NDK version 10.4.0, or r10e, and Qt Creator works with that just fine. I had issues with running a later version of the NDK. Your mileage may vary, as they say.

QNX

QNX is a commercial UNIX-like operating system, which is currently owned by Blackberry. It is not open source, but I thought I would mention it here, as Qt runs on QNX, and is being used commercially in the market.

Connecting to a remote device

This is a book about Qt development and I will stick to Qt Creator. The method is nearly the same to connect to any device, with a few minor differences. You can also connect via **Secure Shell** (**SSH**) and friends with a terminal. I often use both methods, as each has its own advantages and disadvantages.

Qt Creator

I remember when what is now called Qt Creator was first released for internal testing at Nokia. At that time, it was called Workbench. It was basically a good text editor. Since that time, it has gained heaps of awesome features and it is my go to IDE for Qt-based projects.

Qt Creator is a multi-platform IDE, it runs on macOS X, Windows, and Linux. It can connect to Android, iOS, or generic Linux devices. You can even get SDKs for devices such as UBports (Open Source Ubuntu Phone) or Jolla Phones.

To configure your device, in Qt Creator navigate to **Tools** | **Options...** | **Devices** | **Devices**.

Generic Linux

A generic Linux device could be a custom embedded Linux device or even a Raspberry Pi. It should be running an SSH server. Since I used an RPI, I will demonstrate with that.

The following is the devices tab showing connection details for a Raspberry Pi:

As you can see here, the most important item is probably **Host name**. Make sure the IP address in the **Host name** configuration matches the actual IP of the device. Other devices may have a direct USB connection instead of using the regular network.

Android

You will need Android SDK and NDK installed.

Android is a device that uses a direct USB connection, so it will be easier copying the application binary files when it runs the application:

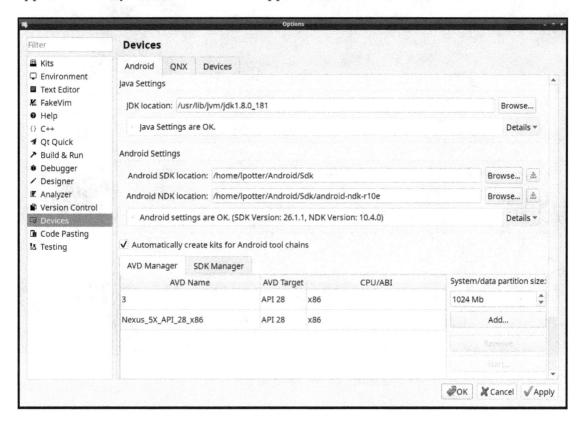

Qt Creator more or less configures this connection automatically.

iOS

Make sure your device is seen by Xcode first, then Qt Creator will automatically pick it up and use it.

It would look similar to this image:

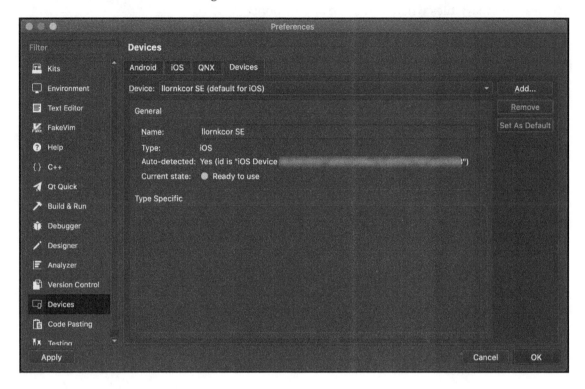

Notice the little green LED-like icon? Ya, all good to go!

Bare metal

If your device does not run an SSH server, you can connect with it using `gdb/gdbserver` or a hardware server. You will need to enable the plugin first. In Qt Creator, navigate to **Help** | **About Plugins** | **Device Support**, and then select **BareMetal**. The bare metal connection uses OpenOCD that you can get from `http://openocd.org`. OpenOCD is not some new anxiety disorder, but an on-chip debugger that runs through a JTAG interface. Qt Creator also has support for the ST-LINK debugger. Both use JTAG connectors. There are USB JTAG connectors as well as the traditional JTAG interface, which do not require any device drivers to get connected.

Writing this section brought back memories of when Trolltech got the Trolltech Greenphone up and running, as well as working on some other devices, like the OpenMoko phone. Good times!

Now that we have a connected device, we can start debugging.

Debugging remotely

Developing software is hard. All software has bugs. Some bugs are more painful than others. The worst kind are probably when you have a random crash that requires a specific sequence of events to trigger that reside on a read-only filesystem remote device that was built in release mode. Been there. Done that. Even got a t-shirt. (I have many Trolltech and Nokia t-shirts left over from days gone by.)

Remote debugging traditionally involves running the gdbserver command on the device. On very small machines where there isn't enough RAM to run gdb directly, running gdbserver on the remote device is probably the only way to use gdb. Let's put on some groove salad and get cracking!

gdbserver

You may want to experience remote debugging without a UI, or something weird like that. This will get you started. The `gdbserver` command needs to be running on the remote device, and there needs to be either a serial or TCP connection.

On the *remote* device, run the following command:

```
gdbserver host:1234 <target> <app args>
```

Using the `host` argument will start `gdbserver` running on port `1234`. You could also attach the debugger on a running application by running the following command:

```
gdbserver host:1234 --attach <pid>
```

`pid` is the process ID of the already running application you are trying to debug, which you could get through running the command `ps`, or top, or similar.

On the *host* device, run the following command:

```
gdb target remote <host ip/name>:1234
```

You will then `issue` commands on the host device through the console that is running `gdb`.

If you run into a crash, after it happens you can type `bt` to get a backtrace listing. If you have a crash memory dump, or core dump as it's called, on the remote, `gdbserver` does not support debugging core memory dumps remotely. You will have to run `gdb` itself on the remote in order to do this.

Using `gdb` via the command line might be fun to some, but I prefer a UI, since it is easier to remember things to be done. Having a GUI that can do remote debugging can help if you are not very familiar with running `gdb` commands, as this can be a daunting task. Qt Creator can do remote debugging, so let's move on to debugging with Qt Creator.

Qt Creator

Qt Creator uses `gdbserver` on the device, so it is essentially just a UI interface. You will need to have Python scripting support for `gdbserver` on the device; otherwise, you will see a message **Selected build of GDB does not support Python scripting**, and it will not work.

For the most part, debugging with Qt Creator works out-of-the-box for Android, iOS, and any supported Boot to Qt device.

Load any project in Qt Creator and it can handle C++ debugging, as well as debugging into Qt Quick projects. Make sure the correct settings are configured in the **Run Settings** page down where it says **Debugger settings** within Qt Creator to enable `qml` debugging and/or C++ debugging if you need it.

Add the following to your project and rebuild:

```
CONFIG+=debug qml_debug
```

Add this to the application startup arguments `-qmljsdebugger=port:<port>`, `host:<ip>`.

To interrupt the execution of the app, click on the icon whose tooltip says '**Interrupt GDB for "yourapp"**'. You can then inspect the value of variables and step though the code.

Set a breakpoint somewhere—right-click on the line in question and select **Set Breakpoint on line** .

Press *F5* to start the application build (if needed). Once successfully built, it will be transferred and executed on the device and the remote debugging service is started. It will, of course, stop execution on your breakpoint, if you have one set. To continue normal execution, press *F5* until you hit that painful crash, and then you can inspect that wonderful backtrace! From here, you can hopefully gather enough clues to fix it.

Other key commands supported by Qt Creator by default are as follows:

- *F5*: start / continue execution
- *F9*: toggle breakpoint
- *F10*: step over
- *Ctrl* + *F10*: run to current line
- *F11*: step into
- *Shift* + *F11*: step out

Let's try it out. Load the source code for this chapter.

To toggle a breakpoint on the current line in the Qt Creator editor, press *F9* on Linux and Windows, or *F8* on macOS as follows:

```
23      ····QString·a("crab");
24      ····QString·b("Apple");
```

Now start the debugger by pressing *F5* to run the app in the debugger. It will stop execution on our line as follows:

```
22
23      ····QString·a("crab");
24      ····QString·b("Apple");
25
26      ····qDebug()·<<·a·<<·b;
```

See that little yellow arrow? It informs us that the execution has stopped on this line, before the statement has been executed.

You will be able to see the following values for the variables:

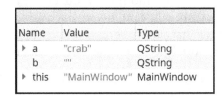

As you can see, the breakpoint stopped execution before QString b has been initialized, so the value is " ". If you push *F10* or step over, the QString b gets initialized and you can see the new value as follows:

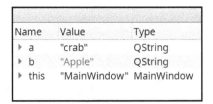

You will notice from the following screenshot that the execution line gets moved to the next statement as well:

```
22
23      ....QString·a("crab");
24      ....QString·b("Apple");
25
26      ....qDebug()·<<·a·<<·b;
27  }
```

You can also edit the breakpoint by right-clicking on the breakpoint in the editor and selecting **Edit Breakpoint**. Let's set a breakpoint on Line 20 in the for loop as follows:

```
19      ....for·(int·i·=·0;·i·<·20;·i++)·{
20      ......qDebug()·<<·i;
21      ....}
```

Right-click and select **Edit Breakpoint** to open the **Edit Breakpoint Properties** dialog as follows:

Edit the **Condition** field and add i == 15 as follows and click **OK**:

Run the app in the debugger by clicking *F5*. Click on the **Strings** button. When it hits the breakpoint, you can see it stopped when **i** contains the value **15**:

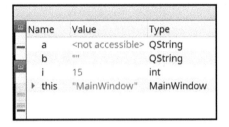

You could then step into, or step over.

Let's now look at a crash bug, which is a divide by zero crash when you push the **crash** button.

Set a breakpoint at line *31*. Run the debugger, and it will stop just before the crash. Now step over. You should see a dialog popup as follows:

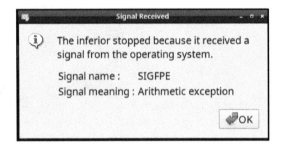

Oh my. Now that is ugly.

In the stack view shown in the following screenshot, you can see where the program has crashed:

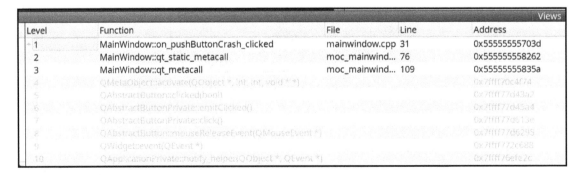

Yep, it is right where I put it! Bad things happen when you divide by zero in C++.

Summary

Debugging is a powerful process and often required to fix bugs. You can either run debuggers such as gdb from the command line, or you should be able to connect the gdb to a debugger server running on a remote device. Running a GUI-based debugger is much more fun. You should be able to debug your Qt app running on your mobile or embedded device from Qt Creator over a remote connection.

The next step is to deploy your application. We will explore various ways to deploy your application on a few different mobile and embedded platforms.

13
Deploying to Mobile and Embedded

Mobile phones, tablets, and watches have their platform ways to deploy apps—usually through an app store. Deploying plugins and other libraries needs special attention. In this chapter, we will discuss alternative OSes, such as Jolla's Sailfish OS, as embedded devices have several options. I use a Raspberry Pi as an example for embedded Linux devices.

For the major mobile phone app stores, you will need to digitally sign your package with a security certificate, which the systems use as a way to identify the author enough to trust the application.

A certificate involves a public-private key pair. The private key is just that. You keep that private. The public certificate is publicly distributable. I won't go into the cryptography involved here. Qt Creator calls these certificates the keystore and you can use Qt Creator to generate these self-signed certificates.

We will examine the following deployment targets:

- For Android
- For iOS
- For alternative OSes
- For embedded Linux

Keep backups of your digital certificates as you will not be able to update your applications in the stores if you lose them.

Deployment for Android

Android does not need Google Play Store to install apps; it's just most convenient. There are other marketplaces to choose from, such as Aptoid, Yandex, F-Droid, and Amazon.

You can also sideload apps. Sideloading is installing an app by transferring the package by USB, memory card, or over the internet, without the use of the official store.

Qt Creator technically can sideload the package of the application you are working on. It can install the package, or simply run the executable on the device without installing it.

Essentially, you can put a package file on your web server, have people download it to their phones or computers, and let them manually install it.

You could also make it available on Google Play Store, by officially publishing it. You need to be able to sign it with a certificate that you get from your developer account. This certificate for Android does not need to be signed by a certificate authority, but can be self-signed.

The package

After developing and testing your application, you will need to make a package so people can install it. There are a few ways to make a package, via Qt Creator or by the command line.

androiddeployqt

The command line tool named `androiddeployqt` comes with Qt Creator for the Android SDK. This is a command line tool to help build and sign an Android package. To view its help output, run `/path/to/androiddeployqt --help`. I am not going to go into the command line deployment other than to say it is available to use.

Qt Creator

Digital certificates in Qt Creator are handled on the **Build Settings** page for your projects. Now, let's build:

1. **Projects** | **Build** will get you to the **Build Settings** page. If you are making a release, make sure your project is in **Release** mode from the top of this page:

2. Navigate to **Build Steps** | **Build Android APK**:

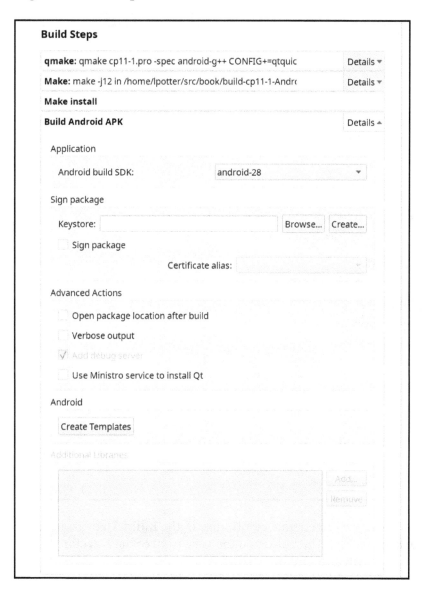

3. You will need to click on **Create Templates**, which will create the `manifest.xml` file needed for the store. The two main entries are **Package name** and **Application icon**. I used Android Studio to create different sized icons, because it was the most efficient at creating several icons at once. I started with a large PNG. Be sure to select the correct **SDK** versions, or it will become an issue later on:

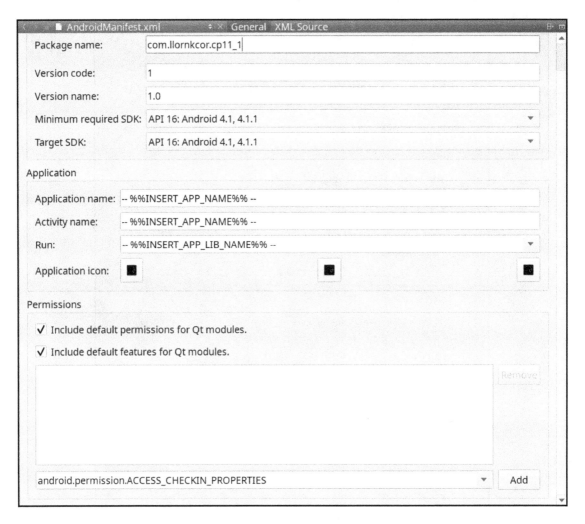

4. You also need to generate certificates. In the **Build Steps** page, find the **Sign package** section. The button on the very right of the **Keystore** entry, which says **Create...**, will bring up this dialog:

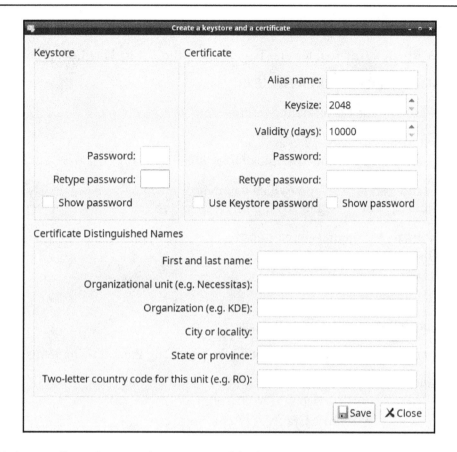

5. You will need to supply a **Password** for both **Keystore** and **Certificate**, which are the private and public keys.
6. You also need to supply your **name**, **Organization**, **City**, **State**, and **Two-letter country code**.

Do another build and it will create a signed Android package that you can install or upload to the store. You are now ready to test your app in a closed, internal, or open test track. Qt Creator will not help you there.

Test track

You can set up an internal test track for testers in your organization. If you are a one-stop shop, you are the tester!

Internal tests

You can have up to 100 internal testers. To create a testers list on Play Console, navigate to:

Settings | **Manage testers** | **CREATE LIST**

You will need to provide the following information:

- **List name**
- **Email addresses** (you can upload a csv email list too!)

You can then add testers to an app (which you may or may not have previously added). Now select your app and navigate to **App releases**, and select one of the following:

- **Internal**: An internal closed track
- **Alpha**: A closed track
- **Beta**: An open track
- **Production**: Release

Click on **Manage (Internal)** | **Create Release**

You need to make sure your account is set up, by providing the following items:

- Store Listing—screenshots
    - Hi-res icon (512 x 512 png)
    - Feature Graphic (1,024 x 500 jpg, png)
- Content rating
- Pricing and distribution

You need to upload an app package before you add testers. Once you upload your package app, you will need to select **Review**.

You will then be asked to select your testers list when you navigate back to the **App releases** page. You should receive a URL to share with your tester that they can use to download and install your internally-tested application package.

That red icon at the top of the page means something needs to be checked.:

In my case, I hadn't actually selected any testers to this release yet.

When you add to the list of testers, it should look similar to this:

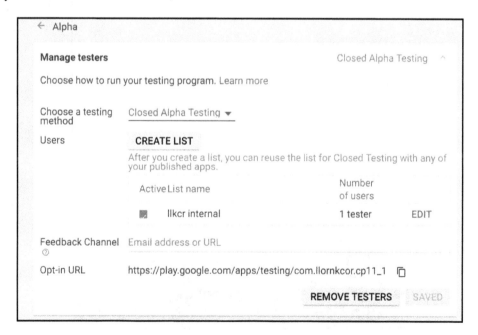

Here is what your testers will see in the **Google Play** Store:

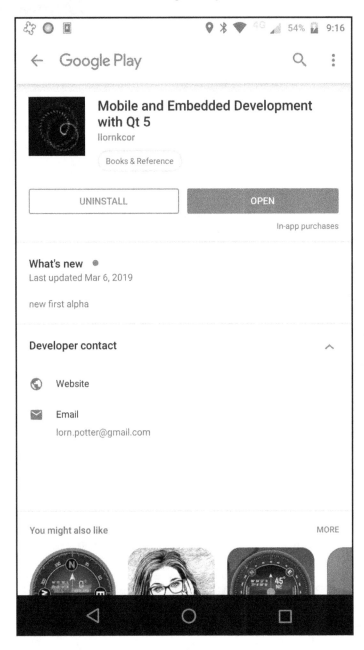

Deployment on iOS is very similar, so let's look at that.

Deployment for iOS

The iOS store is perhaps the most restrictive and complicated of all the mobile app stores to submit an app to. It also has more exhaustive submission guidelines, such as: it should not replicate the functionality of native applications. The process to get to the point of submitting an app is also more complicated.

The package

Qt Creator comes with support for creating and signing iOS packages. As with Android, you will need certificates from your developer account. This is what you see when you log into the developer account at Apple:

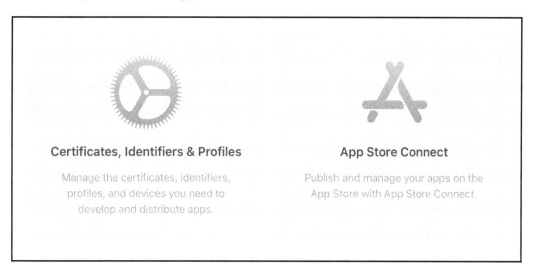

Certificates, Identifiers & Profiles

Manage the certificates, identifiers, profiles, and devices you need to develop and distribute apps.

App Store Connect

Publish and manage your apps on the App Store with App Store Connect.

Once in your developer account, click on the icon labelled **Certificates, Identifiers & Profiles** to add certificates. Notice the list under **Certificates** at the left:

There are two types of **Certificates**: **Development** and **Production**. **Production** certificates are for release distribution. If you do not have **Production** certificates, add one now by clicking on the + icon:

That will open up the following dialog:

Development

○ **iOS App Development**
Sign development versions of your iOS app.

○ **Apple Push Notification service SSL (Sandbox)**
Establish connectivity between your notification server and the Apple Push Notification service sandbox environment to deliver remote notifications to your app. A separate certificate is required for each app you develop.

Production

◉ **App Store and Ad Hoc**
Sign your iOS app for submission to the App Store or for Ad Hoc distribution.

○ **Apple Push Notification service SSL (Sandbox & Production)**
Establish connectivity between your notification server, the Apple Push Notification service sandbox, and production environments to deliver remote notifications to your app. When utilizing HTTP/2, the same certificate can be used to deliver app notifications, update ClockKit complication data, and alert background VoIP apps of incoming activity. A separate certificate is required for each app you distribute.

○ **Pass Type ID Certificate**
Sign and send updates to passes in Wallet.

Select **App Store and Ad Hoc**. **Ad Hoc** means you can only install this on a few testing devices.

Next, under **Identifiers**, select **App IDs**:

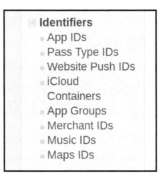

There are two types of **App IDs**:

- *Wildcard*, which can be used for multiple apps if they do not require iCloud or in-app purchases.
- *Explicit*, which is used for in-app purchases per application. You will need one of these for each app you have that uses in-app purchasing.

You also need **Provisioning Profiles**:

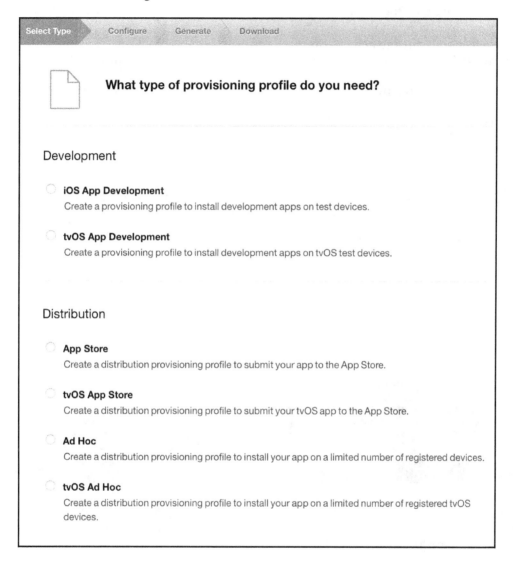

As you can see here, there are different types of provisioning profiles. Under **Distribution**, select **App Store** then the **Continue** button at the bottom of the page. Select the **App ID** you created previously. Select the **Certificate** also previously created. You will need to name this profile, and it must match the **Bundle Identifier** for your package. Download this and put it somewhere you remember; you will need this to sign the app through XCode.

You will also need an App ID, so select **App IDs** from the left and create a new one.

Now we can sign a release mode package.

Qt Creator

Qt Creator cannot create iOS packages itself. We need to use the Xcode project file that Qt Creator generates in order to create the package using Xcode.

Select a **Release** build and then run qmake to create the Xcode project file we will use.

Xcode

From the build directory, open the resulting `<target>.xcodeproj` in Xcode. Select the project from the left to open up the project settings.

Click on **General** and then unselect **Automatically manage signing** to be able to manually select the package signing:

You can import the profile by selecting **Import Profile...** from the drop-down list that is labeled **Provisioning Profile**, and then select **Import Profile....** Once the file dialog opens, you can navigate to where you put the `<profilename>.mobileprovision` file:

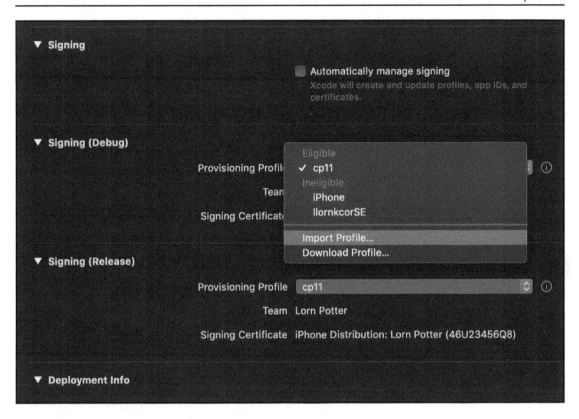

You can do this for both **Signing (Debug)** and **Signing (Release)**.

Your **Bundle Identifier** must match the name of the provisioning profile. If it doesn't, it will alert you:

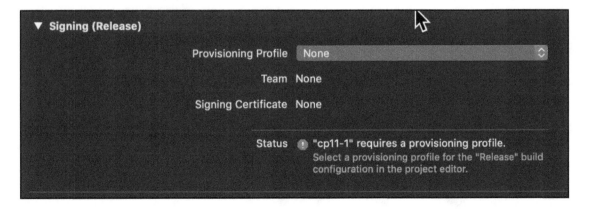

The certificates have been applied:

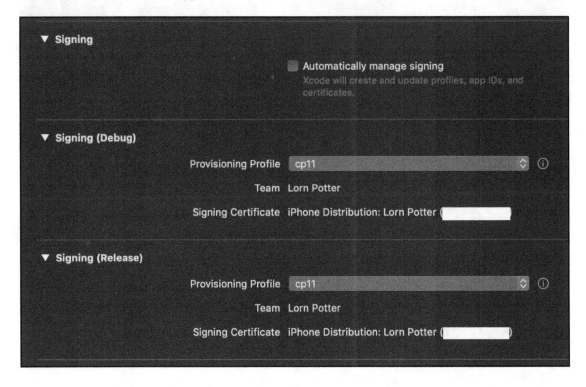

Test and build the package. It should ask you for your administrator password when it signs the package.

To build a release mode package, navigate to **Product** | **Scheme** | **Edit Scheme** | **Info** | **Build Configuration**.

Select **Release** and then build the package.

From here, you need to grab your web browser again and navigate to the **App Store Connect**, select **My Apps**, and click on the + to create a **New App**:

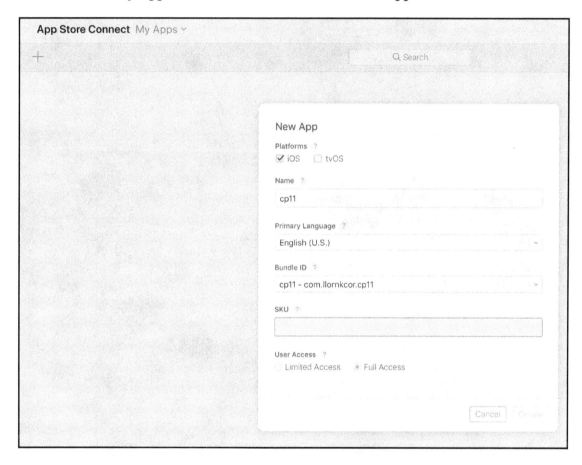

From here, fill out **App Information**, **Pricing, and Availability**.

Next, add screenshots, icons, store text, and other items for the apps page and the App Store.

To upload your app from Xcode, select **Product | Archive.**

This will create the package and open the **Archives** window.

You should validate the package before uploading it to the App Store, so select **Validate App**, then select **profile** and the **certificates**:

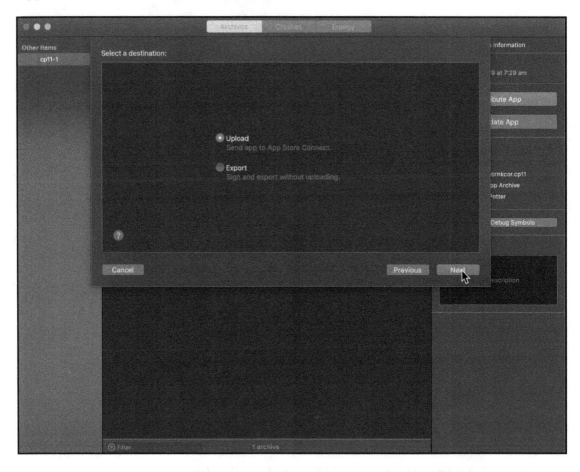

Fix any issues you may have. This also includes adding screenshots, app icons, and so on for the storefront for the app. Then you can click the **Distribute App** button.

Alternative OSes

There are other mobile and embedded OSes available that you may or may not have heard about, such as my favorite alternative mobile OS: Jolla's Sailfish. Another operating system is UBports, which is an Open Source version of Canonical's now-defunct mobile phone OS, Ubuntu Touch.

Sailfish OS

Sailfish OS is a continuation of Nokia's MeeGo, which was a continuation of Maemo.

The UI is developed by Jolla, and the base OS is open source Mer, which is developed by Jolla and the community.

Jolla has an app store they named Harbour (`http://harbour.jolla.com`). At this time, you cannot sell apps through this app store:

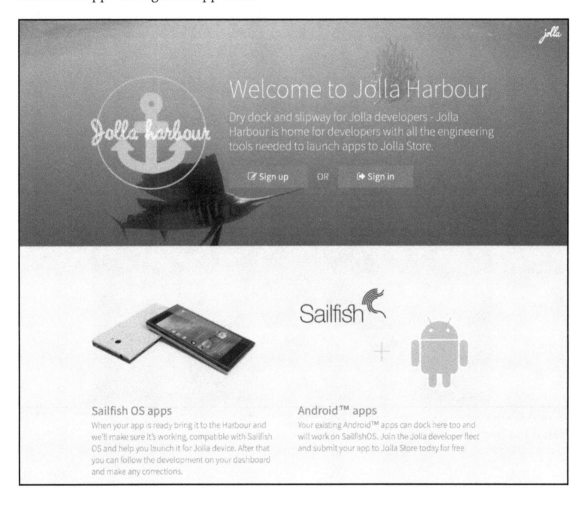

This is what my developer page looks like:

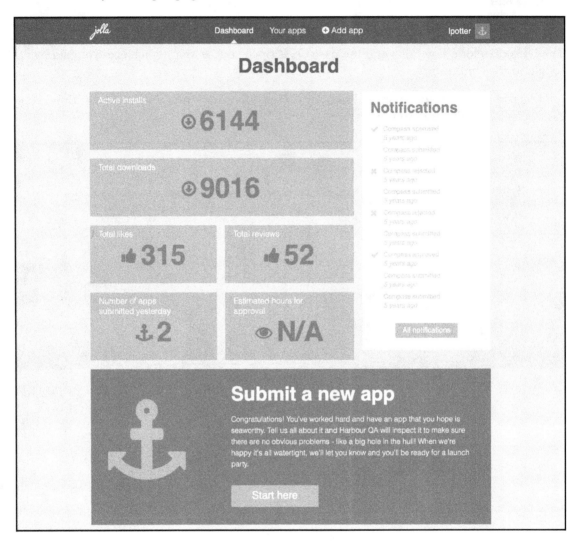

Yes, it has been five years since I have updated it.

You can install Jolla onto certain Android phones—or if you are lucky enough to have actual Jolla hardware or perhaps a phone that comes with Jolla installed, you have access to Harbour through the Store app. Here is the view of the **Top apps** page:

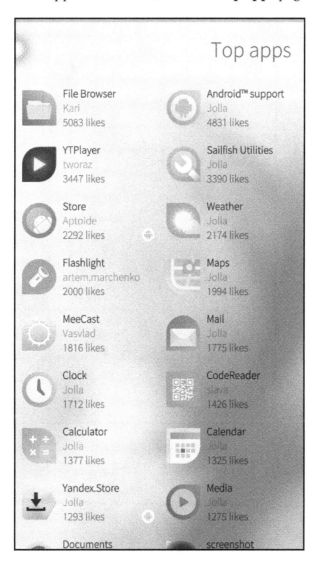

I have an old app in that store named **Compass** that I need to update to the new Sailfish OS version 3:

I need to download the Application SDK from https://releases.sailfishos.org/sdk/ installers/1.24/.

I install it on my Linux development machine as follows:

```
chmod +x SailfishOSSDK-Beta-1.24-Qt5-linux-64-offline.run
./SailfishOSSDK-Beta-1.24-Qt5-linux-64-offline.run
```

The **Sailfish OS SDK** is now installed!

After Qt Creator opens with the Sailfish SDK, click on **Sailfish OS** from the icons on the left:

You should see a message that says that the build engine is not running, so we need to start it. Click on **Start the build engine!**:

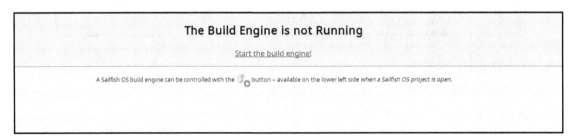

Once you do that, you get into the **Sailfish Control Centre**, where you can add components into the SDK, and apply updates if needed.

Set up your device, if you have one, and navigate to **Tools | Options | Devices.**

Once there, click on **Tools | Options | Kits** and select the **armv7hl** kit.

In the **Devices** options section, make sure that the Jolla device that you set up previously is selected and not the emulator, unless that is what you want to run the app on.

The Jolla SDK build engine runs in a Virtual machine, so it can be used from any platform. The IDE it uses is Qt Creator. Jolla is unique in that you can run native apps for Jolla OS, but also it can run Android Apps. The gotcha about Android support is that there is no Google Play API.

Now that the build engine (cross compiler) is running, we can build our app, test, and then make a package to upload to Harbour.

From the **Projects | Run** settings, make sure either **Deploy by Copying Binaries** or **Deploy As RPM Package** is selected for **Method**. If you are running it without installing a package, select the **Copying** method. Here is my updated **Compass** app running on the Jolla phone:

Once you have built the release package, navigate to **Sailfish OS | Publishing,** click on the **RPM** validator, and select your package file to validate.

The Jolla Store interface is the least complicated of the mobile app stores here, in part because there are no sales of apps.

Click on **Add app**. You will have to fill out the following:

- **Title**
- **Details**:
    - **description**
    - **summary**
    - **recent changes**
- **Categorization**
- **Compatibility**
- **Visual Assts (screenshots, icons)**
- **Contact details**
- **Any message to QA**

Once you submit your application to Jolla's Harbour store, you will see something like this image:

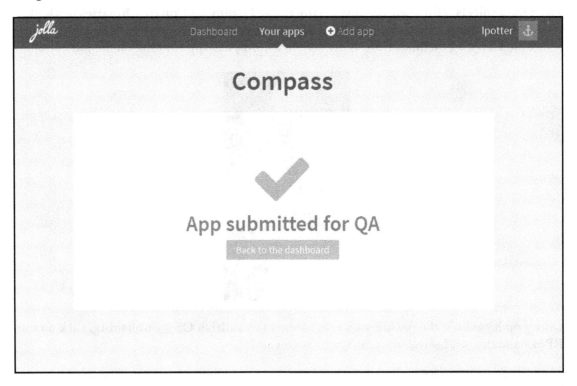

QA will check the package and either reject or accept it.

UBports

UBports is a convergent OS based on Canonical's now-defunct mobile offering, Ubuntu Touch. Convergent means it is designed to run on desktops and mobile devices. It runs on various mobile phones and tablets, and the UI is based on Qt and QML. I run mine on a Nexus 4. I won't go into detail, but I did wanted to mention it. More information can be found on: `https://ubports.com/`

There is a simple installer for putting UBports on a supported device at `https://ubuntu-touch.io/get-ut`.

You can grab the SDK at `https://docs.ubports.com/en/latest/appdev/index.html`.

Clickable

The UBports SDK produces click packages from the command line. The Ubuntu Touch SDK IDE is no longer supported by Canonical or UBports. The store is called OpenStore. You need an account, and the URL to submit apps is `https://open-store.io/submit`.

Embedded Linux

Embedded Linux devices come in many different sizes and varieties. Some may have app stores, but most don't. There are various methods to get the OS and apps on the devices.

OS deployment

The deployment of the operating system will on your device, as some embedded devices have very specific methods in which the operating system is deployed to the device. In the case of Raspberry Pi, it is easy to copy the image to an SD card and put that into the RPI and boot it up.

I have a script named `writeIso` that I use; it consists of two lines:

```
#!/bin/bash
sudo dd if=$1 of=$2 bs=4M status=progress
```

I run it something like this:

```
./writeIso /path/to/deviceImage.img /dev/sdc
```

Other devices may have a `flash` method, whereby the image gets copied directly onto the device. This can be as low-level as having to use JTAG, or it could be higher-level, such as using Android's `adb` command. Sometimes, you have to write the image to an SD card, put that into the device, and then, with some combination of keys or buttons, flash that image into the machine's ROM.

App deployment

With a distribution with a package manager such as Raspbian or Yocto, you can easily distribute your application, either by installing directly on the device or adding to the package repository. In the case of Yocto, you can have a local repository to distribute.

To get the package file onto the device, you can use Qt Creator and set up a generic Linux device. This requires an SSH server running on the device and some kind of network connection.

You can also use the `scp` command to copy packages and/or binaries to the device. This also requires an SSH server.

Summary

There are a few different methods for OS and app deployment. Mobile phones have app stores, which also have various methods to submit apps. Usually, these are done though a web browser. You should be able to publish apps to the Android, iOS, and alternative OS, such as Jolla's Sailfish, app stores.

You should also be able to distribute your app onto embedded devices, such as Raspberry Pi.

In the next chapter, we will traverse into the brand new technology of Qt for WebAssembly that allows Qt applications to run in a web browser.

14
Universal Platform for Mobiles and Embedded Devices

Deploying applications and targeting all the different platforms can take heaps of time and cost thousands of dollars. There's a new target platform for Qt applications called Qt for WebAssembly that allows Qt and Qt Quick apps to be run over a network from a web browser. You will learn how to set up, cross build, deploy, and run Qt applications that work on any device with a modern web browser. You could say that Qt for WebAssembly is the universal platform.

We will detail the following material:

- Technical requirements
- Getting started
- Building with the command line
- Building with Qt Creator
- Deploying for mobile and embedded devices
- Tips, tricks and suggestions

What is this WebAssembly thing?

WebAssembly is neither strictly Web nor Assembly. At the same time, it is a little of both.

At the technical level, it is a new binary instruction format for a stack-based virtual machine, according to the WebAssembly web site at `http://webassembly.org`. It runs in a modern web browser, but people are naturally experimenting with this and it can now run standalone and experimentally like any other app, with support being written for the Linux kernel.

Through the use of the Emscripten tool, it can be compiled from C and C++. Emscripten is a tool written in Python that uses LLVM to transpile C++ code into WebAssembly byte code that can be loaded by a web browser.

WebAssembly byte code runs in the same sandbox as JavaScript, so consequentially it has the same limitations regarding access to the local file system, as well as living in one thread. It also has the same security benefits. Although there is work being done to fully support pthreads, it is, at the time of this writing, still experimental.

Technical requirements

Easy install binary SDK from the following Git repository:

- **Emscripten sdk** `https://github.com/emscripten-core/emscripten.git`

Alternatively, manually compile the SDK. You can download the sources from these Git URLs:

- **Emscripten** `https://github.com/emscripten-core/emscripten.git`
- **Binaryen** `https://github.com/WebAssembly/binaryen.git`
- **LLVM** `https://github.com/llvm/llvm-project.git`

Getting started

According to the Emscripten website at `https://emscripten.org/`:

Emscripten is a toolchain that uses LLVM to transpile code to WebAssembly to run in a web browser at near native speeds.

These are the two ways to install Emscripten:

- Clone the repository, install precompiled binaries
- Clone the repositories, build them

I recommend the first one, as LLVM is very time-consuming to build. It is also recommended to use Linux or macOS. If you are on Windows, you can install the Linux subsystem and use that, or use MinGW compiler. The Visual Studio compiler does not seem to support output targets with the four-letter extensions that Emscripten outputs, namely `.wasm` and `.html`.

Download Emscripten

You need to have Git and Python installed for this—just clone the `emscripten sdk`:

```
git clone https://github.com/emscripten-core/emscripten.git.
```

In there, are Python scripts to help out, the most important one being `emsdk`.

First run `./emsdk --help` to print out some documentation on how to run it.

Then you need to install and then activate the SDK as follows:

```
./emsdk install latest
./emsdk activate latest
```

You can target a specific SDK; you can see what is available by running the following command:

```
./emsdk list
```

Then install a particular version of the SDK by running the following commands:

```
./emsdk install sdk-1.38.16-64bit
./emsdk activate sdk-1.38.16-64bit
```

The `activate` command sets up the `~/.emscripten` file that contains the environment settings needed by Emscripten.

To be able build with it, you need to source the `emsdk_env.sh` file as follows:

```
source ~/emsdk/emsdk_env.sh
```

Qt targets a certain Emscripten version that is known to be good for that version. For Qt 5.11, Qt for WebAssembly has its own branch—`wip/webassembly`. It has been integrated into 5.12 as a tech preview, and in 5.13 for official support. At the time of this writing, it is planned to be included with Qt Creator as a binary install.

Building an Emscripten SDK manually

If you want to build Emscripten manually, such as to compile upstream LLVM which has support for transpiling directly to WebAssembly binary instead of writing first to JavaScript and then to WebAssembly. This can speed up compile times, but, at the time of this writing, is still experimental. This makes use of adding an argument to the linker -s WASM_OBJECT_FILES=1.

For more information on using WASM_OBJECT_FILES, see https://github.com/emscripten-core/emscripten/issues/6830.

Technical requirements

You will need to install node.js and cmake packages from your OS. Clone the following resources:

```
mkdir emsdks
cd emsdks
git clone -b 1.38.27 https://github.com/kripken/emscripten.git
git clone -b 1.38.27 https://github.com/WebAssembly/binaryen.git
git clone https://github.com/llvm/llvm-project.git
```

Emscripten does not have to be built, as it is written in Python.

To build binaryen, enter the following code:

```
cd binaryen
cmake .
make
```

To build LLVM, enter the following code:

```
mkdir llvm
cmake ../llvm-project/llvm -
DLLVM_ENABLE_PROJECTS="clang;libcxx;libcxxabi;lld" -
DCMAKE_BUILD_TYPE=Release -DLLVM_TARGETS_TO_BUILD=WebAssembly -
DLLVM_EXPERIMENTAL_TARGETS_TO_BUILD=WebAssembly

make
```

Run emscripten to write the configure file as follows:

```
cd emscripten
./emcc --help
```

This will create a ~/.emscripten file. Copy this file over to your emsdks directory as follows:

```
cp ~/.emscripten /path/to/emsdks/.emscripten-vanillallvm
```

To set up the environment, write a script as follows:

```
#!/bin/bash
SET EMSDK=/path/to/emscripten
SET LLVM=/path/to/llvm/bin
SET BINARYEN=/path/to/binaryen
SET PATH=%EMSDK%;%PATH%
SET EM_CONFIG=/path/to/emsdks/.emscripten-vanillallvm
SET EM_CACHE=/path/to/esdks/.emscripten-vanillallvm_cache
```

Save it somewhere as `emsdk-env.sh`.

You will need to make this executable, so run `chmod +x emsdk-env.sh`.

Whenever you need to set up the build environment, simply run this script and use the same console to build.

Now that we are ready, let's see how to configure and build Qt.

Configuring and compiling Qt

You can find information on Qt for WebAssembly at this URL: `https://wiki.qt.io/Qt_for_WebAssembly`

I guess we need the sources. You can get them through Qt Creator, or you can `git clone` the repository. Using Git, you have more control over which version and any branch if needed.

For 5.12 and 5.13, you can simply clone the following tag:

`git clone http://code.qt.io/qt/qtbase.git -b v5.12.1`

Alternatively, you can `clone` this tag:

`git clone http://code.qt.io/qt/qtbase.git -b v5.13.0`

As with any new technology, it is moving fast, so grab the latest version you can. For this book, we are using Qt 5.12, but I included mentioning other versions as they have many bug fixes and optimizations.

Now we can configure and compile Qt!

For 5.12 and 5.13 it was simplified to the following:

```
configure -xplatform wasm-emscripten -nomake examples -nomake tests
```

If you need threads, 5.13 has support for multithreading WebAssembly, but you also need to configure the browser to support it.

Once it configures, all you need to do is run make!

Then, to build your Qt app for running in a web browser, simply use the `qmake` command from the build directory and run it on your apps pro file. Not every Qt feature is supported—like local filesystem access and threads. `QOpenGLWidget` is also not supported, although `QOpenGLWindow` works fine. Let's see how to build using then command line.

Building with the command line

Building a Qt for a WebAssembly application requires you to source the Emscripten environment file, so run this in your console command as follows:

```
source /path/to/emsdk/emsdk_env.sh
```

You will need to add the path to Qt for WebAssembly `qmake` as follows:

```
export PATH=/path/to/QtForWebAssembly/bin:$PATH.
```

Of course, you must replace `/path/to` with the actual filesystem path.

You are then ready for action! You build it just like any other Qt app, by running `qmake` as follows:

```
qmake mycoolapp.pro && make.
```

If you need to debug, rerun `qmake` with `CONFIG+=debug` as follows:

```
qmake CONFIG+=debug mycoolapp.pro && make.
```

This will add various Emscripten specific arguments to the compiler and linker.

Once it is built, you can run it by using the `emrun` command from Emscripten, which will start a simple web server and serve the `<target>.html` file. This will, in turn, load up `qtloader.js`, which, in turn, loads up the `<target>.js` file, which loads the `<target>.wasm` binary file:

```
emrun --browser firefox --hostname 10.0.0.4 <target>.html.
```

You can also give `emrun` the directory, such as:

```
emrun --browser firefox --hostname 10.0.0.4 ..
```

This gives you time to bring up the browser's web console for debugging. Now, let's see how to use Qt creator for building.

Building with Qt Creator

It is possible to use Qt Creator to build and run your Qt app once you have compiled Qt itself from the command line.

The Build environment

In Qt Creator, navigate to **Tools | Options... | Kits**

Then go to the **Compilers** tab. You need to add emcc as a C compiler, and em++ as a C++ compiler, so click on the **Add** button and select **Custom** from the drop-down list.

First select **C** and add the following details:

- **Name:** emcc (1.38.16)
- **Compiler path:** /home/user/emsdk/emscripten/1.38.16/emcc
- **Make path:** /usr/bin/make
- **ABI:** x86 linux unknown elf 64bit
- **Qt mkspecs:** wasm-emscripten

Select **C++** and add the following details:

- **Name:** emc++(1.38.16)
- **Compiler path:** /home/user/emsdk/emscripten/1.38.16/em++
- **Make path:** /usr/bin/make
- **ABI:** x86 linux unknown elf 64bit
- **Qt mkspecs:** wasm-emscripten

Click **Apply**.

Go to the tab labeled **Qt Versions** and click on the **Add** button. Navigate to where you build Qt for WebAssembly, and, in the bin directory, select the **qmake**. Click **Apply**.

Go to the tab labeled **Kits**, and click on the **Add** button. Add the following details:

- **Name**: `Qt %{Qt:Version} (qt5-wasm)`
- **Compiler**: `C: emcc (1.38.16`
- **Compile**: `C++: em++ (1.38.16)`
- **Qt version**: `Qt (qt5-wasm)`

The Run environment

You need to make your application's project active to build for Qt for WebAssembly. From the left-hand side buttons in Qt Creator, select **Projects**, and select your **Qt for WebAssembly** kit.

Running the WebAssembly apps in Qt Creator is currently a bit tricky, as you need to specify `emrun` as a custom executable and then the build directory or `<target>.html` file as its argument. You can also specify which browser to run. You can run Chrome using the `--browser chrome` argument.

To get a list of found browsers, run the command `emrun --list_browsers`.

You can even run the app in an Android device that is connected to a USB using the `--android` argument. You need to have **Android Debug Bridge** (**adb**) command installed and running.

Anyway, now that we know how to run the app, we need to tell the Qt Creator project to run it.

Go to **Projects** | **Run**. In the **Run** section, select **Add** | **Custom Executable** and add the following details:

- **Executable**: `/home/user/emsdk/emrun <target>.html`
- **Working directory**: `%{buildDir}`

Now we are ready to build and run. Here is how it should look:

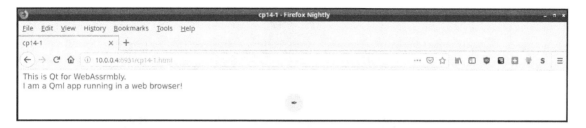

We can even run OpenGL apps! Here is the `hellogles3` example from Qt running in the Firefox browser on Android:

We can also run declarative apps! Here is Qt Quick's `qopenglunderqml` example app:

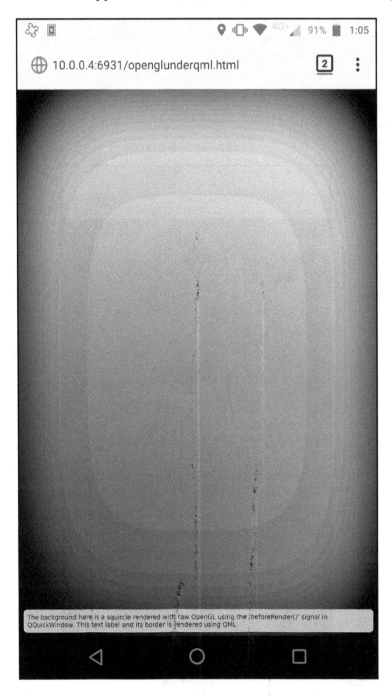

Deploying for mobile and embedded devices

Really, deploying for mobile and embedded devices is only copying the resulting files from Emscripten built onto a CORS-enabled web server.

Any web browser that supports WebAssembly will be able to run it.

Of course, there are considerations regarding screen size.

For testing, you can run your application using the emrun command from the Emscripten SDK. If you plan on testing from another device other than localhost, you will need to use the --hostname argument to set the IP address that it uses.

There are Python scripts for CORS-enabled web servers for testing as well. The Apache web server can also be configured to support CORS.

There are five files that currently need to be deployed—qtloader.js, qtlogo.svg, <target>.html, <target>.js, and <target>.wasm. The .wasm file is the big WebAssembly binary, statically linked. Following are few suggestions to help you along with the process.

Tips, tricks, and suggestions

Qt for WebAssembly is treated by Qt as a cross platform build. It is an emerging technology and, as such, some features required may need special settings configuration to be changed or enabled. There are a few things you need to keep in mind when using it as a target.

Here, I run through some tips regarding Qt for WebAssembly.

Browsers

All major browsers now have support for loading WebAssembly. Firefox seems to load fastest, although Chrome has a configuration that can be set to speed it up (look at chrome://flags for #enable-webassembly-baseline). Mobile browsers that come with Android and iOS also work, although these may run into out of memory errors, depending on the application being run.

Qt 5.13 for WebAssembly has added experimental support for threads, which rely onSharedArrayBuffer support in the browsers. This has been turned off by default, due to Spectre vulnerabilities, and need to be enabled in the browsers.

In Chrome, navigate to `chrome://flags` and enable `#enable-webassembly-threads`.

In Firefox, navigate to `about://config` and enable `javascript.options.shared.memory`.

Debugging

Debugging is done by using the debugging console in the web browser. Extra debugging capabilities can be enabled by invoking `qmake` with `CONFIG+=debug`, even with a Qt compiled in release mode. Here is what a crash can look like:

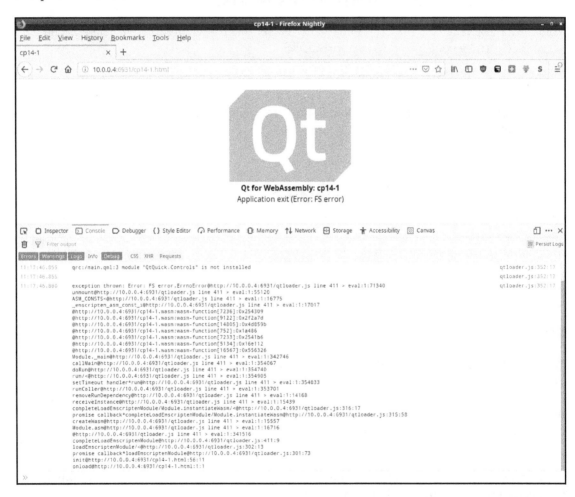

You can also remote debug from your phone and see the remote browser's JavaScript console output on your desktop. See the following link:

```
https://developer.mozilla.org/en-US/docs/Tools/Remote_Debugging
```

Networking

Simple download requests can be made with the usual `QNetworkAccessManager`. These will go through `XMLNetworkRequest`, and will require a CORS-enabled server to download from. Typical `QTCPSocket` and `QUdpSockets` get transpiled into WebSockets. Your web server needs to support WebSockets, or you can use the Websockify tool, which is available from the following link:

```
https://github.com/novnc/websockify
```

Fonts and filesystem access

System fonts cannot be accessed, and must be included and embedded into the application. Qt embeds one font.

Filesystem access is also not currently supported, but will be in the future by using a Qt WebAssembly specific API.

OpenGL

OpenGL is supported as OpenGL ES2, which gets transpiled into WebGL.

There are a few differences between OpenGL ES2 and WebGL that you should be aware of if you plan on using OpenGL in your WebAssembly application. WebGL is more strict generally.

Here are some of the differences for WebGL:

- A buffer may only be bound to one `ARRAY_BUFFER` or `ELEMENT_ARRAY_BUFFER` in it's lifetime
- No client side `Arrays`
- No binary shaders, `ShaderBinary`
- Enforces `offset` for `drawElements`; `vertexAttribPointer` and `stride` arguments for `vertexAttribPointer` are a multiple of the size of the data type

- `drawArrays` and `drawElements` are restricted from requesting data outside the bounds of a buffer
- Adds `DEPTH_STENCIL_ATTACHMENT` and `DEPTH_STENCIL`
- `texImage2D` and `texSubImage2D` size based on the `TexImageSource` object
- `copyTexImage2D`, `copyTexSubImage2D` and `readPixels` cannot touch pixels outside of `framebuffer`
- Stencil testing and bound `framebuffer` have restricted drawing
- `vertexAttribPointer` value must not exceed the value 255
- `zNear` cannot be greater than `zFar`
- constant color and constant alpha cannot be used with `blendFunc`
- no support for `GL_FIXED`
- `compressedTexImage2D` and `compressedTexSubImage2D` are not supported
- GLSL token size limited to 256 characters
- GLSL is ASCII characters only
- GLSL limited to 1 level of nested structures
- Uniform and attribute location lengths limited to 256 characters
- `INFO_LOG_LENGTH`, `SHADER_SOURCE_LENGTH`, `ACTIVE_UNIFORM_MAX_LENGTH`, and `ACTIVE_ATTRIBUTE_MAX_LENGTH` have been removed.
- Texture type passed to `texSubImage2D` must match `texImage2D`
- Calls that `read` and `write` to same texture (feedback loop) not allowed
- Reading data from missing attachment is not allowed
- Attribute aliasing not allowed
- `gl_Position` initial value defined as (0,0,0,0)

For more information, see the following web pages:

WebGL 1.0 `https://www.khronos.org/registry/webgl/specs/latest/1.0/#6`

WebGL 2.0 `https://www.khronos.org/registry/webgl/specs/latest/2.0/#5`

Supported Qt modules

Qt for WebAssembly supports the following Qt modules:

- qtbase
- qtdeclarative
- qtquickcontrols2
- qtwebsockets
- qtmqtt
- qtsvg
- qtcharts
- qtgraphicaleffects

Other caveats

Secondary event loops do not work in Qt for Webassembly. This is because the Emscripten event loop it needs to tie in to does not return. If you need to pop up a dialog, do not call exec() but call show(), and use signals to get a return value.

Virtual keyboards on mobile platforms like Android and iOS do not automatically pop up. You can use Qt Virtual Keyboard directly in your project.

Summary

Qt for WebAssembly is a new and upcoming platform for Qt, which runs Qt apps in a web browser.

You should now be able to download or build the Emscripten SDK, and use to build Qt for WebAssembly. You can now run Qt apps in a web browser, including mobile and embedded devices, as long as the browser supports WebAssembly.

In the final chapter, we explore building a complete Linux embedded operating system.

Building a Linux System

15

Building your own Linux system for use on an embedded device can be an overwhelming task. Knowing what software is needed to get the stack up and running; knowing what the software dependencies are; finding the software to download and downloading it; configuring, building, and packaging all of that software—it could literally take weeks of time. It used to back in the good old days. Now, there are some great tools to facilitate building a custom Linux filesystem. You can be up and running on an embedded device within a day, if you have a machine powerful enough.

Prototyping is always the first step in device creation. Having the correct tools will streamline this process. Embedded systems need to boot fast and directly into a Qt application, such as an automotive instrument cluster. In this chapter, you will learn about creating a full software stack for embedded Linux systems using Yocto and Boot to Qt for Device Creation. A Raspberry Pi device will be used as a target to demonstrate how to build the operating system.

We will be looking at the following:

- Bootcamp – Boot to Qt
- Rolling your own – custom embedded Linux
- Deploying to an embedded system

Bootcamp – Boot to Qt

We have already discussed Qt Company's Boot to Qt system in Chapter 12, *Cross Compiling and Remote Debugging*. Provided with Boot to Qt are configuration files for you to use to create a custom operating system. It requires the BitBake software and the Yocto Project, which is an open source project to help to build custom Linux-based systems, which itself is based on my old friend, OpenEmbedded.

There is a script named `b2qt-init-build-env` in
the `/path/to/install/dir/<Qtversion>/Boot2Qt/sources/meta-boot2qt/b2qt-init-build-env` file of this book that will initialize the build for a Raspberry Pi. You run
the command from a build directory of your choice.

To get a list of supported devices, use the `list-devices` argument. The output on my
system is as follows:

```
~/development/builds/boot2qt $> $~/Qt-comm/5.11.2/Boot2Qt/sources/meta-boot2qt/b2qt-init-build-env list-devices
Available device configurations:
  am335x-evm
  apalis-imx6
  apalis-imx8
  beagleboard
  beaglebone
  colibri-imx6
  colibri-imx7
  colibri-vf
  emulator
  h3ulcb
  imx6dlsabresd
  imx6qsabresd
  imx7dsabresd
  imx7s-warp
  imx8qmlpddr4arm2
  intel-corei7-64
  jetson-tx1
  jetson-tx2
  m3ulcb
  nitrogen6x
  nitrogen7
  raspberrypi0
  raspberrypi2
  raspberrypi3
  raspberrypi
  salvator-x
  smarc-samx6i
  tegra-t18x
```

You will need to initialize the build system and build environment, so run the script
named `b2qt-init-build-env`, which is in the directory that Boot to Qt is installed:

```
/path/to/install/dir/<Qtversion>/Boot2Qt/sources/meta-boot2qt/b2qt-init-build-env init --device raspberrypi3
```

Replace `/path/to/install/dir` with the directory path that `Boot2Qt` is in, typically
`~/Qt`. Also, replace `<Qtversion>` to whatever Qt version is installed there. If you are using
a different device, change `raspberrypi3` to one that is in the list of supported devices by
`Boot2Qt`.

Yocto comes with scripts and configurations so that you can build your own system and
customize it, and perhaps add MySQL database support. The B2Q script, `setup-environment.sh`, will help to set up the environment for development.

You need to export your device type into the MACHINE environmental variable and source the environment setup script:

```
export MACHINE=raspberrypi3
source ./setup-environment.sh
```

Now, you can build the default image by using the following command:

```
bitbake b2qt-embedded-qt5-image
```

You could first customize it by adding a package that you need that isn't there by default—let's say the mysql plugin, so that we can access a database remotely! Let's look at how we can do that.

Rolling your own – custom embedded Linux

Yocto has a history and got its start from the OpenEmbedded project. The OpenEmbedded project got its name in the programming world from the OpenZaurus project. At that time, I was involved with OpenZaurus and projects surrounding that, with the original focus being the Sharp Zaurus that ran Trolltech's Qtopia using a different operating system. OpenZaurus was an open source replacement OS that users could flash onto their devices. The evolution of the build system went from being the Makefile-based Buildroot to being displaced by BitBake.

You can, of course, build Poky or Yocto for this section. I am going to use the Boot2Qt configurations.

To get started with Yocto so that you can customize it, make a base image by using the following command:

```
bitbake core-image-minimal
```

This will take quite a bit of time.

The basic customization procedure would be the same as customizing Boot to Qt, regarding adding layers and recipes, as well as customizing already existing recipes.

System customization

By default, the Boot2Qt rpi image does not contain the MySQL Qt plugin, so the MySQL example I mentioned previously will not work. I added it by customizing the image build.

Yocto and all BitBake-derived systems use a `conf/local.conf` file so that you can customize the image build. If you do not have one already, after you run `setup-environment.sh file`, create a `local.conf` file and add the following line of code:

```
PACKAGECONFIG_append_pn-qtbase = " sql-mysql"
```

The `sql-mysql` part comes from Qt's configure arguments, so this is telling `bitbake` to add the `-sql-mysql` argument to the configure arguments, which will build the MySQL plugin and hence include it in the system image. There are other options, but you will need to look in `meta-qt5/recipes-qt/qt5/qtbase_git.bb` and see the lines that start with `PACKAGECONFIG`.

There is one other customization I need to do, which has nothing to do with Qt. OpenEmbedded uses the `www.example.com` URL to test for connectivity. For whatever reason, my ISP's DNS does not have an entry for `https://www.example.com`, so I initially could not reach it, and the build failed straight away. I could have added a new DNS to my computer's network configuration, but it was faster to tell `bitbake` to use another server for its online check, so I added the following line to my `conf/local.conf` file:

```
CONNECTIVITY_CHECK_URIS ?= "https://www.google.com/"
```

If you need more extensive customization, you can create your own `bitbake` layer, which is a collection of recipes.

local.conf file

The `conf/local.conf` file is where you can make local changes to the image build. Like `PACKAGECONFIG_append_pn-`, which we mentioned in the previous section, there are other ways to add packages and issue other configuration commands. The templated `local.conf` has loads of comments to guide you in the process.

`IMAGE_INSTALL_append` allows you to add packages into the image.

`PACKAGECONFIG_append_pn-<package>` allows you to append package-specific configurations to the package. In the case of `qtbase`, it allows you to add arguments to the configure process. Each recipe will have specific configurations.

meta-<layer> directories

Layers are a way to add packages or add functionality to existing packages. To create your own layer, you will need to create a template directory structure in the sources/ directory, where you initialized the bitbake build. Change <layer> to whatever name you are going to use:

```
sources/meta-<layer>/
sources/meta-<layer>/licenses/
sources/meta-<layer>/recipes/
sources/meta-<layer>/conf/layer.conf
sources/meta-<layer>/README
```

The licenses/ directory is where you put any license files for the package.

Any recipes you may add go into recipes/ directly. There's more on this a bit later.

The layer.conf file is the controlling configuration for the layer. A place to start with this file could be as follows, filled in with generic entries:

```
BBPATH .= ":${LAYERDIR}"
BBFILES += "${LAYERDIR}/recipes-*/*/*.bb \
${LAYERDIR}/recipes-*/*/*.bbappend"
BBFILE_COLLECTIONS += "meta-custom"
BBFILE_PATTERN_meta-custom = "^${LAYERDIR}/"
BBFILE_PRIORITY_meta-custom = "5"
LAYERVERSION_meta-custom = "1"
LICENSE_PATH += "${LAYERDIR}/licenses"
```

Change meta-custom to whatever you want to name it.

Once you have created the layer, you will need to add it to the conf/bblayers.conf file, which is in the directory that you initialized in the Boot2Qt build. In my case, this was ~/development/b2qt/build-raspberrypi3/.

We can now add one or more packages to our custom layer.

<recipe>.bb files

You can also create your own recipe if you have existing code, or if there is a software project somewhere that you want to include in the system image.

In the custom layer, we created a recipes/ directory where our new recipe can live.

To get a feel of how recipes can be written, take a look at some of the recipes included with Boot2Qt or Yocto.

There are some scripts that can help in the creation of recipe files, that is, `devtool` and `recipetool`. The `devtool` and `recipetool` commands are fairly similar in what they do. `Devtool` makes it easier if you need to apply patches and work on the code. Sometimes, your software needs to be developed or debugged on the actual device, for example, if you are developing something that uses any sensors. `Devtool` can also build the recipe so that you can work the kinks out.

devtool command

The output of `devtool --help` is as follows:

```
~/development/b2qt/build-raspberrypi3 $> $devtool --help
usage: devtool [--basepath BASEPATH] [--bbpath BBPATH] [-d] [-q]
               [--color COLOR] [-h]
               <subcommand> ...

OpenEmbedded development tool

options:
   --basepath BASEPATH  Base directory of SDK / build directory
   --bbpath BBPATH      Explicitly specify the BBPATH, rather than getting it
                        from the metadata
   -d, --debug          Enable debug output
   -q, --quiet          Print only errors
   --color COLOR        Colorize output (where COLOR is auto, always, never)
   -h, --help           show this help message and exit

subcommands:
  Beginning work on a recipe:
    add                 Add a new recipe
    modify              Modify the source for an existing recipe
    upgrade             Upgrade an existing recipe
  Getting information:
    status              Show workspace status
    search              Search available recipes
  Working on a recipe in the workspace:
    build               Build a recipe
    rename              Rename a recipe file in the workspace
    edit-recipe         Edit a recipe file in your workspace
    configure-help      Get help on configure script options
    update-recipe       Apply changes from external source tree to recipe
    reset               Remove a recipe from your workspace
    finish              Finish working on a recipe in your workspace
  Testing changes on target:
    deploy-target       Deploy recipe output files to live target machine
    undeploy-target     Undeploy recipe output files in live target machine
    build-image         Build image including workspace recipe packages
  Advanced:
    create-workspace    Set up workspace in an alternative location
    extract             Extract the source for an existing recipe
    sync                Synchronize the source tree for an existing recipe
Use devtool <subcommand> --help to get help on a specific command
```

The most important arguments would be `add`, `modify,` and `upgrade`.

For `devtool`, I will use a Git repository URL to add my repository of `sensors-examples`:

> **devtool add sensors-examples https://github.com/lpotter/sensors-examples.git**

Running the preceding command will output something similar to this:

```
~/development/b2qt/build-raspberrypi3 $> $devtool add sensors-examples https://github.com/lpotter/sensors-examples.git
WARNING: Host distribution "ubuntu-18.04" has not been validated with this version of the build system; you may possibly experience unexpected failur
es. It is recommended that you use a tested distribution.
WARNING: Host distribution "ubuntu-18.04" has not been validated with this version of the build system; you may possibly experience unexpected failur
es. It is recommended that you use a tested distribution.
WARNING: Host distribution "ubuntu-18.04" has not been validated with this version of the build system; you may possibly experience unexpected failur
es. It is recommended that you use a tested distribution.
Loading cache: 100% |#######################################################################################################| Time: 0:00:00
Loaded 2888 entries from dependency cache.
Parsing recipes: 100% |#####################################################################################################| Time: 0:00:00
Parsing of 2109 .bb files complete (2108 cached, 1 parsed). 2889 targets, 350 skipped, 2 masked, 0 errors.

Summary: There was 1 WARNING message shown.
NOTE: Fetching git://github.com/lpotter/sensors-examples.git;protocol=https...
NOTE: Using default source tree path /home/lpotter/development/b2qt/build-raspberrypi3/workspace/sources/sensors-examples
NOTE: Recipe /home/lpotter/development/b2qt/build-raspberrypi3/workspace/recipes/sensors-examples/sensors-examples_git.bb has been automatically crea
ted; further editing may be required to make it fully functional
```

We need to try and build the package to see whether it succeeds or fails, which we can do by running the following command:

> **devtool build sensors-examples**

We might need to edit this `.bb` file to make it build if it does fail.

In the case of `sensors-examples`, we will get the following output:

```
~/development/b2qt/build-raspberrypi3 $> $devtool build sensors-examples
WARNING: Host distribution "ubuntu-18.04" has not been validated with this version of the build system; you may possibly experience unexpected failur
es. It is recommended that you use a tested distribution.
Loading cache: 100% |################################################################################################| Time: 0:00:01
Loaded 2888 entries from dependency cache.
Parsing recipes: 100% |##############################################################################################| Time: 0:00:00
Parsing of 2109 .bb files complete (2108 cached, 1 parsed). 2889 targets, 350 skipped, 2 masked, 0 errors.
NOTE: Resolving any missing task queue dependencies

Build Configuration:
BB_VERSION           = "1.34.0"
BUILD_SYS            = "x86_64-linux"
NATIVELSBSTRING      = "universal"
TARGET_SYS           = "arm-poky-linux-gnueabi"
MACHINE              = "raspberrypi3"
DISTRO               = "b2qt"
DISTRO_VERSION       = "2.3.4"
TUNE_FEATURES        = "arm armv7ve vfp thumb neon vfpv4 callconvention-hard cortexa7"
TARGET_FPU           = "hard"
SDKMACHINE           = "x86_64"
meta
meta-poky
meta-raspberrypi
meta-oe
meta-python
meta-networking
meta-initramfs
meta-multimedia
meta-boot2qt
meta-boot2qt-distro
meta-raspberrypi-extras
meta-mingw
meta-qt5
workspace            = "<unknown>:<unknown>"

Initialising tasks: 100% |###########################################################################################| Time: 0:00:02
NOTE: Executing SetScene Tasks
NOTE: Executing RunQueue Tasks
sensors-examples-1.0+git999-r0 do_compile: NOTE: sensors-examples: compiling from external source tree /home/lpotter/development/b2qt/build-raspberry
pi3/workspace/sources/sensors-examples
NOTE: Tasks Summary: Attempted 1287 tasks of which 1279 didn't need to be rerun and all succeeded.

Summary: There was 1 WARNING message shown.
```

We've built it!

You will find this build in `tmp/work/cortexa7hf-neon-vfpv4-poky-linux-gnueabi/sensors-examples/1.0+git999-r0`.

If you want to edit a recipe, you can use `devtool` and then create a patch so that you can use it:

```
devtool modify qtsensors
```

We get the following output by running the preceding command:

This will duplicate the recipe in your local workspace so that you can edit without losing it when you update bitbake. Now, you can edit the sources, in this case, of qtsensors. I have a patch to add a qtsensors plugin for the Raspberry Pi Sense HAT, so I am going to manually apply that now:

My patch is old, and I need to fix it up. You can build this on its own by running the following command:

```
devtool build qtsensors
```

This initially fails to find `rtimulib.h`, so we need to add a dependency on that lib.

In the OpenEmbedded Layer Index, there is a `python-rtimu` recipe, but it does not export the headers or build the library, so I will create a new recipe based on the Git repository, as follows:

```
devtool add rtimulib https://github.com/RPi-Distro/RTIMULib.git
```

This is a `cmake`-based project, and I will need to modify the recipe to add some `cmake` arguments. To edit this, I can simply run the following command:

```
devtool edit-recipe rtimulib
```

I added the following lines, which use `EXTRA_OECMAKE` to disable some of the demos that depend on Qt 4. I think at one time, I had a patch that ported it to Qt 5, but I cannot find it. The last `EXTRA_OEMAKE` tells `cmake` to build in the Linux directory. Then, we tell `bitbake` it needs to inherit the `cmake` stuff:

```
EXTRA_OECMAKE = "-DBUILD_GL=OFF"
EXTRA_OECMAKE += "-DBUILD_DRIVE=OFF"
EXTRA_OECMAKE += "-DBUILD_DRIVE10=OFF"
EXTRA_OECMAKE += "-DBUILD_DEMO=OFF"
EXTRA_OECMAKE += "-DBUILD_DEMOGL=OFF"
EXTRA_OECMAKE += "Linux"
inherit  cmake
```

We then need to edit our `qtsensors_git.bb` file so that we can add a dependency on this new package. This will allow it to find the headers:

```
DEPENDS += "rtimulib"
```

When I run the build command, `bitbake qtsensors`, it will make sure my `rtimulib` package is built, then apply my `sensehat` patch to `qtsensors`, and then build and package that up!

`Recipetool` is another way that you can create a new recipe. It is simpler in design and usage than `devtool`.

recipetool command

The output of using the `recipetool create --help` command is as follows:

```
~/development/b2qt/build-raspberrypi3 $> $recipetool create --help
usage: recipetool create [-h] [-o OUTFILE] [-m] [-x EXTRACTPATH] [-N NAME]
                         [-V VERSION] [-b] [--also-native]
                         [--src-subdir SUBDIR] [-a] [--keep-temp]
                         [--fetch-dev]
                         source

Creates a new recipe from a source tree

arguments:
  source                Path or URL to source

options:
  -h, --help            show this help message and exit
  -o OUTFILE, --outfile OUTFILE
                        Specify filename for recipe to create
  -m, --machine         Make recipe machine-specific as opposed to
                        architecture-specific
  -x EXTRACTPATH, --extract-to EXTRACTPATH
                        Assuming source is a URL, fetch it and extract it to
                        the directory specified as EXTRACTPATH
  -N NAME, --name NAME  Name to use within recipe (PN)
  -V VERSION, --version VERSION
                        Version to use within recipe (PV)
  -b, --binary          Treat the source tree as something that should be
                        installed verbatim (no compilation, same directory
                        structure)
  --also-native         Also add native variant (i.e. support building recipe
                        for the build host as well as the target machine)
  --src-subdir SUBDIR   Specify subdirectory within source tree to use
  -a, --autorev        When fetching from a git repository, set SRCREV in the
                        recipe to a floating revision instead of fixed
  --keep-temp           Keep temporary directory (for debugging)
  --fetch-dev           For npm, also fetch devDependencies
```

As an example, I ran `recipetool -d create -o rotationtray_1.bb`
`https://github.com/lpotter/rotationtray.git`:

```
~/development/b2qt/build-raspberrypi3 $> $recipetool -d create -o rotationtray_1.bb https://github.com/lpotter/rotationtray.git
DEBUG: Found bitbake path: /home/lpotter/development/b2qt/sources/poky/bitbake
DEBUG: Loading plugins from /home/lpotter/development/b2qt/sources/poky/meta-poky/lib/recipetool...
DEBUG: Loading plugins from /home/lpotter/development/b2qt/build-raspberrypi3/lib/recipetool...
DEBUG: Loading plugins from /home/lpotter/development/b2qt/sources/poky/meta/lib/recipetool...
DEBUG: Loading plugins from /home/lpotter/development/b2qt/sources/meta-raspberrypi/lib/recipetool...
DEBUG: Loading plugins from /home/lpotter/development/b2qt/sources/meta-openembedded/meta-oe/lib/recipetool...
DEBUG: Loading plugins from /home/lpotter/development/b2qt/sources/meta-openembedded/meta-python/lib/recipetool...
DEBUG: Loading plugins from /home/lpotter/development/b2qt/sources/meta-openembedded/meta-networking/lib/recipetool...
DEBUG: Loading plugins from /home/lpotter/development/b2qt/sources/meta-openembedded/meta-initramfs/lib/recipetool...
DEBUG: Loading plugins from /home/lpotter/development/b2qt/sources/meta-openembedded/meta-multimedia/lib/recipetool...
DEBUG: Loading plugins from /home/lpotter/development/b2qt/sources/meta-boot2qt/meta-boot2qt/lib/recipetool...
DEBUG: Loading plugins from /home/lpotter/development/b2qt/sources/meta-boot2qt/meta-boot2qt-distro/lib/recipetool...
DEBUG: Loading plugins from /home/lpotter/development/b2qt/sources/meta-boot2qt/meta-raspberrypi-extras/lib/recipetool...
DEBUG: Loading plugins from /home/lpotter/development/b2qt/sources/meta-mingw/lib/recipetool...
DEBUG: Loading plugins from /home/lpotter/development/b2qt/sources/meta-qt5/lib/recipetool...
DEBUG: Loading plugin create_qt5
DEBUG: Loading plugins from /home/lpotter/development/b2qt/sources/poky/scripts/lib/recipetool...
DEBUG: Loading plugin create_npm
DEBUG: Loading plugin create_buildsys
DEBUG: Loading plugin create_buildsys_python
DEBUG: Loading plugin create
DEBUG: Loading plugin create_kernel
DEBUG: Loading plugin newappend
DEBUG: Loading plugin setvar
DEBUG: Loading plugin append
DEBUG: Loading plugin create_kmod
WARNING: Host distribution "ubuntu-18.04" has not been validated with this version of the build system; you may possibly experience unexpected failu
res. It is recommended that you use a tested distribution.
WARNING: Host distribution "ubuntu-18.04" has not been validated with this version of the build system; you may possibly experience unexpected failu
res. It is recommended that you use a tested distribution.
WARNING: Host distribution "ubuntu-18.04" has not been validated with this version of the build system; you may possibly experience unexpected failu
res. It is recommended that you use a tested distribution.
Loading cache: 100% |#########################################################################################################| Time: 0:00:00
Loaded 2888 entries from dependency cache.

Summary: There was 1 WARNING message shown.
NOTE: Fetching git://github.com/lpotter/rotationtray.git;protocol=https...
DEBUG: Fetcher accessed the network with the command git -c core.fsyncobjectfiles=0 ls-remote https://github.com/lpotter/rotationtray.git
```

Using the `-d argument` means it will be more verbose, so I excluded some output:

```
DEBUG: Loading recipe handlers
DEBUG: Handler: KernelRecipeHandler (priority 100)
DEBUG: Handler: PythonRecipeHandler (priority 70)
DEBUG: Handler: NpmRecipeHandler (priority 60)
DEBUG: Handler: CmakeRecipeHandler (priority 50)
DEBUG: Handler: AutotoolsRecipeHandler (priority 40)
DEBUG: Handler: SconsRecipeHandler (priority 30)
DEBUG: Handler: Qmake5RecipeHandler (priority 21)
DEBUG: Handler: QmakeRecipeHandler (priority 20)
DEBUG: Handler: KernelModuleRecipeHandler (priority 15)
DEBUG: Handler: MakefileRecipeHandler (priority 10)
DEBUG: Handler: VersionFileRecipeHandler (priority -1)
DEBUG: Handler: SpecFileRecipeHandler (priority -1)
NOTE: Recipe rotationtray_1.bb has been created; further editing may be required to make it fully functional
```

Now, you may want to edit the resulting file. A great way to learn about `bitbake` is to look at other recipes and see how they do things.

bitbake-layers

OpenEmbedded comes with a script called `bitbake-layers`, which you can use to get information on layers that are available. You can also use this to add a new layer or remove one from the configuration file.

Running `bblayers --help` will give us the following output:

```
~/development/b2qt/build-raspberrypi3 $> $bitbake-layers --help
usage: bitbake-layers [-d] [-q] [--color COLOR] [-h] <subcommand> ...

BitBake layers utility

optional arguments:
  -d, --debug           Enable debug output
  -q, --quiet           Print only errors
  --color COLOR         Colorize output (where COLOR is auto, always, never)
  -h, --help            show this help message and exit

subcommands:
  <subcommand>
    add-layer           Add a layer to bblayers.conf.
    remove-layer        Remove a layer from bblayers.conf.
    flatten             flatten layer configuration into a separate output
                        directory.
    show-layers         show current configured layers.
    show-overlayed      list overlayed recipes (where the same recipe exists
                        in another layer)
    show-recipes        list available recipes, showing the layer they are
                        provided by
    show-appends        list bbappend files and recipe files they apply to
    show-cross-depends  Show dependencies between recipes that cross layer
                        boundaries.
    layerindex-fetch    Fetches a layer from a layer index along with its
                        dependent layers, and adds them to conf/bblayers.conf.
    layerindex-show-depends
                        Find layer dependencies from layer index.

Use bitbake-layers <subcommand> --help to get help on a specific command
```

Running `bitbake-layers show-recipes` will dump all of the available recipes. This list can be quite extensive.

yocto-layer

Yocto has a script named `yocto-layer`, which will create an empty layer directory structure that you can then add with `bitbake-layers`. You can also add an example recipe and a `bbappend` file.

To create a new layer, run `yocto-layer` with the `create` argument:

```
yocto-layer create mylayer
```

This will run interactively and ask you a few questions. I told it yes to both questions to create examples:

```
~/development/b2qt/build-raspberrypi3/workspace/sources $> $yocto-layer create mylayer
Please enter the layer priority you'd like to use for the layer: [default: 6] 98
Would you like to have an example recipe created? (y/n) [default: n] y
Please enter the name you'd like to use for your example recipe: [default: example] myexample
Would you like to have an example bbappend file created? (y/n) [default: n] y
Please enter the name you'd like to use for your bbappend file: [default: example] myexample
Please enter the version number you'd like to use for your bbappend file (this should match the recipe you're appending to): [default: 0.1]

New layer created in meta-mylayer.

Don't forget to add it to your BBLAYERS (for details see meta-mylayer/README).
```

You will then see a new directory tree named `meta-mylayer`. You can then make the new layer available to `bitbake` using `bitbake-layers`, as follows:

```
bitbake-layers add-layer meta-mylayer
```

Use the following command to see the new layer running:

```
bitbake-layers show-layers
```

bbappend files

When I imported the `qtsensors` recipe into my workspace, I could have used a `bbappend` file. When you import the recipe into your workspace, it is essentially being duplicated. Please note, however, that you will no longer be able to build it with `devtool`.

I also mentioned that the `yocto-layer` script can create an example `bbappend` file with a patch so that we can see how it works. The filename that you choose must match whatever recipe you are modifying. The only difference in name would be the extension, which is `.bbappend` for a `bbappend` file.

In the `workspace/conf/local.conf` file, there is a line about BBFILES that tells me where it is looking for `bbappend` files. Of course, you can put them anywhere as long as you tell `bitbake` where they are. Mine is configured for them to be in `${LAYERDIR}/appends/*.bbappend`.

Ours is simple—it only applies a patch. With only the following few lines in the `bbappend` file, we can get it up and running:

- `SUMMARY`: Simple string explaining the patch
- `FILESEXTRAPATHS_prepend`: String of the path to the patch files
- `SRC_URI`: The URL string to the patch file

If we wanted to create a `bbappend` file to patch `qtsensors` with the `sensehat` patch, it would be a four line edit, along with the actual patch. The simple `bbappends` file would look like this:

```
SUMMARY = "Sensehat plugin for qtsensors"
DEPENDS += "rtimulib"
FILESEXTRAPATHS_prepend := "${THISDIR}:"
SRC_URI += "file://0001-Add-sensehat-plugin.patch"
```

It's good practice to place the patch into a directory, but this one is on the same level as the `bbappend` file.

We would need to remove the `qtsensors` recipe that was imported from the workspace before we can build our appended recipe:

```
devtool reset qtsensors
```

Place `qtsensors_git.bbappend` and the patch file into the `appends` directory. To build it, simply run the following command:

```
bitbake qtsensors
```

Now that we can customize our OpenEmbedded/Yocto image, we can deploy to the device.

Deploying to an embedded system

We've built a custom system image, and there are a few different ways that systems can be deployed onto a device. Usually, an embedded device has a particular way of doing this. The image can be deployed to a Raspberry Pi by writing the system image file directly onto a storage disk using dd or similar. Other devices might need to be deployed by writing the filesystem to a formatted disk, or even as low level as using JTAG.

OpenEmbedded

If you plan on using Qt with OpenEmbedded, you should be aware of the `meta-qt5-extra` layer, which contains desktop environments such as LXQt and even KDE5. I personally use both environments and switch back and forth between the two on my desktop, but I prefer LXQt most of the time as it's lightweight.

Building an OpenEmbedded image with LXQt is fairly straightforward, and similar to building a Boot to Qt image.

To see the image targets that are available, you can run the following command:

```
bitbake-layers show-recipes | grep image
```

If you have Boot to Qt, you should see the `b2qt-embedded-qt5-image` layer, which we will use to create the image for Raspberry Pi. You should also see OpenEmbedded's `core-image-base` and `core-image-x11`, which may also be interesting.

 There are other layers available that you can search for and download from `https://layers.openembedded.org/layerindex/branch/master/layers/`.

The deployment method really depends on your target device. Let's see how we can deploy the system image to a Raspberry Pi.

Raspberry Pi

The example in this section targets the Raspberry Pi. You may have a different device, and the process here might be similar.

If you intend to only create a cross `toolchain` that you can use in Qt Creator, you can run the following command:

```
bitbake meta-toolchain-b2qt-embedded-qt5-sdk
```

To create the system image to copy to an SD card, run the following command:

```
bitbake b2qt-embedded-qt5-image
```

The `b2qt-embedded-qt5-image` target will also create the SDK if it is needed. When you let that run for a day or so, you'll have a freshly baked Qt image! I would suggest using the fastest machine you have with the most memory and storage, as a full distro build can take many hours, even on a fast machine.

You can then take the system image and use the device's flash procedure or whatever method it uses to make the filesystem. For the RPI, you put the micro SD into a USB reader, and then run the `dd` command to write the image file.

The resulting file I need to write to the SD card was at the following location:

```
tmp/deploy/images/raspberrypi3/b2qt-embedded-qt5-image-
raspberrypi3-20190224202855.rootfs.rpi-sdimg
```

To write this to the SD card, use the following command:

```
sudo dd if=/path/to/sdimg of=/dev/path/to/usb/drive bs=4M status=progress
```

My exact command was as follows:

```
sudo dd if=tmp/deploy/images/raspberrypi3/b2qt-embedded-qt5-image-
raspberrypi3-20190224202855.rootfs.rpi-sdimg of=/dev/sde bs=4M
status=progress
```

Now, wait until everything has been written to the disk. Plop it into the Raspberry Pi SD slot, power it on, and then you're on your way!

Summary

In this chapter, we learned about how to use `bitbake` to build a custom system image, starting with Qt's Boot to Qt configuration files. The process is similar to building Yocto, Poky, or Ångström. We also learned how use `devtool` to customize Qt's build to add more functionality. Then, we discussed how to add your own recipe using `recipetool` into the image. By doing this, you were also able to add this recipe into a new layer. We finished off by deploying the image onto an SD card so that it could be run on the Raspberry Pi.

Other Books You May Enjoy

If you enjoyed this book, you may be interested in these other books by Packt:

Mastering Qt 5
Guillaume Lazar

ISBN: 9781786467126

- Create stunning UIs with Qt Widget and Qt Quick
- Develop powerful, cross-platform applications with the Qt framework
- Design GUIs with the Qt Designer and build a library in it for UI preview
- Handle user interaction with the Qt signal/slot mechanism in C++
- Prepare a cross-platform project to host a third-party library
- Build a Qt application using the OpenCV API
- Use the Qt Animation framework to display stunning effects
- Deploy mobile apps with Qt and embedded platforms

Qt 5 Projects
Marco Piccolino

ISBN: 9781788293884

- Learn the basics of modern Qt application development
- Develop solid and maintainable applications with BDD, TDD, and Qt Test
- Master the latest UI technologies and know when to use them: Qt Quick, Controls 2, Qt 3D and Charts
- Build a desktop UI with Widgets and the Designer
- Translate your user interfaces with QTranslator and Linguist
- Get familiar with multimedia components to handle visual input and output
- Explore data manipulation and transfer: the model/view framework, JSON, Bluetooth, and network I/O
- Take advantage of existing web technologies and UI components with WebEngine

Leave a review - let other readers know what you think

Please share your thoughts on this book with others by leaving a review on the site that you bought it from. If you purchased the book from Amazon, please leave us an honest review on this book's Amazon page. This is vital so that other potential readers can see and use your unbiased opinion to make purchasing decisions, we can understand what our customers think about our products, and our authors can see your feedback on the title that they have worked with Packt to create. It will only take a few minutes of your time, but is valuable to other potential customers, our authors, and Packt. Thank you!

Index